T0215117

Computer Vision Projects with PyTorch

Design and Develop Production-Grade Models

Akshay Kulkarni
Adarsha Shivananda
Nitin Ranjan Sharma

Apress®

Computer Vision Projects with PyTorch: Design and Develop Production-Grade Models

Akshay Kulkarni
Bangalore, Karnataka, India

Adarsha Shivananda
Hosanagara tq, Shimoga dt, Karnataka, India

Nitin Ranjan Sharma
Bangalore, India

ISBN-13 (pbk): 978-1-4842-8272-4   ISBN-13 (electronic): 978-1-4842-8273-1
https://doi.org/10.1007/978-1-4842-8273-1

Managing Director, Apress Media LLC: Welmoed Spahr
Acquisitions Editor: Celestin Suresh John
Development Editor: Laura Berendson
Coordinating Editor: Shrikant Vishwakarma
Copy Editor: Kezia Endsley

Cover designed by eStudioCalamar

Cover image designed by Freepik (www.freepik.com)

Distributed to the book trade worldwide by Springer Science+Business Media New York, 1 New York Plaza, Suite 4600, New York, NY 10004-1562, USA. Phone 1-800-SPRINGER, fax (201) 348-4505, e-mail orders-ny@springer-sbm.com, or visit www.springeronline.com. Apress Media, LLC is a California LLC and the sole member (owner) is Springer Science + Business Media Finance Inc (SSBM Finance Inc). SSBM Finance Inc is a **Delaware** corporation.

For information on translations, please e-mail booktranslations@springernature.com; for reprint, paperback, or audio rights, please e-mail bookpermissions@springernature.com.

Apress titles may be purchased in bulk for academic, corporate, or promotional use. eBook versions and licenses are also available for most titles. For more information, reference our Print and eBook Bulk Sales web page at http://www.apress.com/bulk-sales.

Any source code or other supplementary material referenced by the author in this book is available to readers on GitHub via the book's product page, located at www.apress.com/ 978-1-4842-8272-4. For more detailed information, please visit http://www.apress.com/source-code.

Printed on acid-free paper

To our families.

Table of Contents

About the Authors

Akshay R Kulkarni is an AI and machine learning (ML) evangelist and a thought leader. He has consulted with several Fortune 500 and global enterprises to drive AI and data science-led strategic transformations. He is a Google developer, author, and regular speaker at major AI and data science conferences (including *Strata, O'Reilly AI Conf,* and *GIDS*). He has been a visiting faculty member for some of the top graduate institutes in India. In 2019, he was featured as one of the top 40 under 40 data scientists in India. In his spare time, he enjoys reading, writing, coding, and helping aspiring data scientists. He lives in Bangalore with his family.

Adarsha Shivananda is a data science and ML Ops leader. He is currently working on creating world-class ML Ops capabilities to ensure continuous value delivery from AI. He aims to build a pool of exceptional data scientists within and outside of the organization to solve problems through training programs, and he strives to stay ahead of the curve. He has worked extensively in the pharmaceutical, healthcare, CPG, retail, and marketing domains. He lives in Bangalore and loves to read and teach data science.

Nitin Ranjan Sharma is a manager at Novartis. He leads a team that develops products using multi-modal techniques. As a consultant, he has developed solutions for Fortune 500 companies and has been involved in solving complex business problems using machine learning and deep learning frameworks. His major focus area and core expertise is computer vision, including solving challenging business problems dealing with images and video data. Before Novartis, he was part of the data science team at Publicis Sapient, EY, and TekSystems Global Services. He is a regular speaker at data science community meetups and an open-source contributor. He also enjoys training and mentoring data science enthusiasts.

About the Technical Reviewer

Jalem Raj Rohit is a senior data scientist at Episource, where he leads all things computer vision. He co-founded ML communities like Pydata Delhi and Pydata Mumbai and organizes and speaks at meetups and conferences.

He has authored two books and a video lesson on the Julia language and serverless engineering. His areas of interest are computer vision, ML Ops, and distributed systems.

Introduction

This book explores various popular methodologies in the field of computer vision in order to unravel its mysteries. We use the PyTorch framework, because it's used by researchers, developers, and beginners to leverage the power of deep learning. This book explores multiple computer vision problems and shows you how to solve them. You can expect an introduction to some of the most critical challenges with hands-on code in PyTorch, which is suitable for beginner and intermediate Python users, along with various methodologies used to solve those business problems.

Production-grade code related to important concepts we present over the course of the book will help you get started quickly. These code snippets can be run on local systems, with or without GPUs (Graphics Processing Units) or on a cloud platform.

We'll introduce you to the concepts of image processing in stages, starting with the basic concepts of computer vision in the first chapter. We'll also delve into the field of deep learning and explain how models are developed for vision-related tasks. You'll get a quick introduction to PyTorch to prepare you for the example business challenges we'll be presenting later in the book. We explore concepts of the revolutionary convolutional neural networks, as well as architectures such as VGG, ResNet, YOLO, Inception, R-CNN, and many others.

The book dives deep into business problems related to image classification, object detection, and segmentation. We explore the concepts of super-resolution and GAN architectures, which are used in many industries. You learn about image similarity and pose estimation, which help with unsupervised problem sets. There are topics related to video analytics, which will help you develop the mindset of using the image and time-based concepts of frames. Adding to the list, the book ends by discussing how these deep learning models can be explained to your business partners. This book aims to be a complete suite for those pursuing computer vision business problems.

CHAPTER 1

The Building Blocks of Computer Vision

Humans have been part of a natural evolutionary pattern for centuries. According to the Flynn Effect, an average person born in recent times has a higher IQ than the average person born in the previous century. Human intelligence allows us to learn, decide, and make new decisions based on our learnings. We use IQ scores to quantify human intelligence, but what about machines? Machines are also part of this evolutionary journey. How have we moved our focus to machines and made them intelligent, as we know them today? Let's take a quick look at this history.

A breakthrough came in the 1940s when programmable digital computers became available, followed by the concept of the *Turing test,* which could measure the intelligence of machines. The concept of the *perceptron* goes back to 1958, when it was introduced as a powerful logical unit that could learn and predict. The perceptron is equivalent to a biological neuron that helps humans function. The 1970s saw fast growth in the field of artificial intelligence, and it has increased exponentially since that time.

Artificial intelligence is the intelligence showcased by a machine, more often when it is trained on historical events. Humans have been trained and conditioned their whole lives. We know, for example, that going too near a fire will cause us to be burned, which is painful and bad for our skin.

© Akshay Kulkarni, Adarsha Shivananda, and Nitin Ranjan Sharma 2022
A. Kulkarni et al., *Computer Vision Projects with PyTorch*,
https://doi.org/10.1007/978-1-4842-8273-1_1

Similarly, a system can be trained to make distinctions between fire and water, based on the features or on historical evidence. Human intelligence is being replicated by machines, which gives rise to what we know as artificial intelligence.

Artificial intelligence encompasses machine learning and deep learning. Machine learning can be thought of as mathematical models that help algorithms learn from data/historical events and formulate decision-making processes. The machine learns the pattern of the data and enables the algorithms to create a self-sustaining system. Its performance can be limiting, such as in the case of huge complex data, which is where deep learning comes into the picture. Deep learning is another subset of artificial intelligence. It uses the concept of the perception, expands it to neural networks, and helps the algorithms learn from various complex data. Even though we have many modeling techniques at our disposal, it's best to find good and explainable results from the simplest of techniques, as stated by *Occam's razor* (the simplest answer is often the best).

Now that we have explored a bit of the history, let's browse through the applicable fields. There are two fields—Natural Language Processing (NLP) and Computer Vision (CV)—that use an immense amount of deep learning techniques to help solve problems. NLP caters to the problem sets defined by our language, essentially one of the most important modes of communication. CV, on the other hand, addresses vision-related problems. The world is full of data that humans can decipher with their senses. This includes vision from the eyes, smell from the nose, audio waves from the ear, taste from the tongue, and sensations from the skin. Using this sensory input, the connected neurons in our brains parse and process the information to make decisions on how to react. Computer vision is one field that addresses the visual side of the machine-learning problem.

This book takes you through the fundamentals and gives you a working knowledge of computer vision.

What Is Computer Vision

Computer vision deals with specific problem sets that rely on images and videos. It tries to decipher the information in the images/videos in order to make meaningful decisions. Just like humans parse an image or a series of images placed sequentially and make decisions about them, CV helps machines interpret and understand visual data. This includes object detection, image classification, image restoration, scene-to-text generation, super-resolution, video analysis, and image tracking. Each of these problems is important in its own way. Studying vision-related problems has gained a lot of attraction after the power of parallel computing came into play.

Applications

The applications of computer vision vary with respect to the industry being discussed. The following sections look at a few of these tasks.

Classification

The simplest form of an image-related problem involving a decision-making process is a supervised technique, called *classification*. Classification simply involves assigning classes to different images. The process can be as simple as an image having one class or it can more complex, when there are multiple classes within the same image. See Figures 1-1 and 1-2.

Figure 1-1. The class in this case is a cat

Figure 1-2. The class in this case is a dog

We can separate the content of such images based on whether they
have an image of a cat or a dog. This is an example of how our eyes
perceive differences. The background of the object we are trying to
classify does not matter, so we need to make sure that it doesn't matter
in the algorithms as well. For example, if we included a logo of some car
company in front of all the dog images, the image classifier network might
learn to classify dogs based on that logo and use it as a shortcut. We later
describe in detail how to incorporate this information into the model.
Classification can be used to identify objects in a production line of a
manufacturing unit.

Object Detection and Localization

An interesting problem that is often encountered is the need to locate a
particular image inside another image and even detect what that might be.
Let's say there is a crowd of people and some are wearing masks and some
are not. We can use a vision algorithm to learn the features of masks, then
use that information to locate a mask relative to the image and detect the
masks. See Figures 1-3 and 1-4.

Figure 1-3. *Class: No mask detected*

Figure 1-4. *Class: A mask is detected in the image*

This analysis can be helpful in detecting license plates of moving vehicles from traffic cameras. Sometimes, due to the resolution of the cameras and the moving traffic, the picture quality is not that great. Super-resolution is one technique that is sometimes used to enhance an image's quality and help identify the numbers on the plate.

Image Segmentation

This process is used to determine edges, curves, and gradients of similar objects placed together in order to separate different objects in an image. A classic unsupervised technique can be used here without the worry of finding good-quality, labeled data. The processed data can further be used as an input to an object detector. See Figure 1-5.

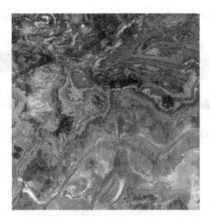

Figure 1-5. *Separating terrains in topological maps*

Anomaly Detection

Another classic, unsupervised way of determining changes is to compare an image to the usual, expected patterns from some training data. Anomaly detection can be used, for example, to determine imperfections in steel pipes when compared to training data. If the machine finds something odd, it will detect an anomaly and inform the line engineers to take care of it. See Figures 1-6a and 1-6b.

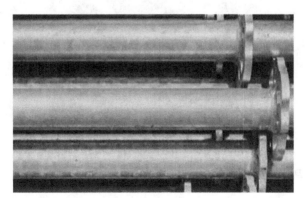

Figure 1-6a. *Perfect examples of steel pipes*

Figure 1-6b. *Anomalies showing up in the pipes*

Video Analysis

There are a lot of use cases for video or sequences of images. The task of object detection on running images can help with CCTV footage. It can also be used to detect abnormalities within the frames per video section.

We will be going through all of these applications in detail in the upcoming chapters. Before that, let's go through a few of the intrinsic concepts that lay the foundation for further understanding computer vision.

Channels

Figure 1-7. *Playing musicians*

One of the most basic and quintessential ideas around computer vision is the *channel*. Think of music being played with multiple instruments; we hear a combination of all the instruments playing together, which essentially constitutes the music in stereo (see Figure 1-7). If we break the music into single components, we can break the sound wave into individual sounds coming from the electric guitar, the acoustic guitar, the piano, and the vocals. After breaking the music into its components, we can modulate each component to get the desired music. There can be an infinite number of combinations if we learn all the musical modulations.

255	255	255	255
255	255	255	255
255	255	255	255
255	255	255	255

Figure 1-8a. *Pixel values corresponding to a white picture*

Figure 1-8b. *Sample white page*

2	36	40	200
195	190	20	180
40	54	200	200
30	40	200	180

Figure 1-8c. *Representational pixel values corresponding to a sketch*

Figure 1-8d. *Sample sketch represented in one channel*

We can extrapolate these concepts to images, which we can break into components of colors. Pixels are the smallest containers of colors. If we zoom in on any digital image, we see small boxes (pixels), which make up the image. The general range of any pixel in terms of the intensity of a channel is 0-255, which is also the range defined by eight bits. Consider Figure 1-8b. We have a white page. If we convert that page to an array, it will give us a matrix of all 255 pixels, as shown in Figure 1-8a. On the other hand, Figure 1-8d, when converted to a matrix, will also have only one channel, and the intensity will be defined by the numbers in the range of 0-255, as depicted in Figure 1-8c. Closer to 0 gives us black and closer to 255 gives us white.

Let's consider a color image. We can break any full-color image into a combination of three primary components (channels)—red, green, and blue. We can break down any color image into some definite combination of red, green, and blue. Thus, RGB (red, green, and blue) becomes the channels of the colored image.

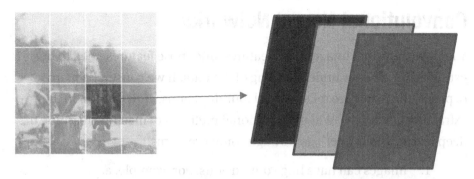

Figure 1-9. *Sample image to blue (left)=0, green (middle)=1, and red (right)=2*

The image in Figure 1-9 can be split into RGB, with the first channel being blue, then green, followed by red. Each pixel in the image can be a certain combination of RGB.

We are not restricted to using RGB as the color channels. We also have HSV (Hue, Saturation, and Value), LAB format, and CMYK (Cyan, Magenta, Yellow, and Black), which are a few representations of channels of an image. Color is a feature and its container is a channel, so every image is made of edges and gradients. We can create any image in the world with just edges and gradients. If you zoom in on a small circle, it should look like a combination of multiple edges and straight lines.

To summarize, channels can be called in as the container for a feature. The features can be the smallest individual characteristics of the image. Color channels are a specific example of channels. Since edges can be features, a channel that only caters to edges can be a channel too. Food for thought—if you were to create a model that would identify a cat or a dog, does the color of the animal affect the model behavior as much as edges and gradients can?

Convolutional Neural Networks

You now know that images have features, and those features need to be extracted for a better understanding of the data. If we consider a matrix of pixels, the pixels are related in all four directions. How do you do the extraction efficiently? Will the traditional methods of machine learning or deep learning help? Let's go through some problems:

1. Images can have huge dimensions. For example, a 2MP image, if it is allowed to capture a 1600x1200 image, will have 1.9 million pixels per image.

2. If we are capturing the data via the images, the data is not always centrally aligned. For example, a cat can be in one corner of one image, and on the next image, it can be in the center. The model should be able to capture the spatial changes in the information.

3. A cat in an image can be rotated along the vertical dimension or the horizontal dimension and still it remains a cat. Thus, we need a robust solution to capture such differences.

We need major upgrades from our regular tabular data approach. If we can break down a problem into smaller manageable pieces, anything can be solved. Herein, we use *convolutional neural networks.* We try to break the image into several feature maps via kernels and use those in sequence to build a model that can then be used for any downstream or pretext task.

Kernels are feature extractors. Features can be edges, gradients, patterns, or any of the small features discussed earlier in this chapter. A square matrix is generally used to operate a convolution task in the image on the first step and on feature maps from the second step. The convolution tasks performed by the kernels can be thought of as the simplest tasks in a dot product. See Figures 1-10a and 1-10b.

2	3	4
2	3	2
3	4	1

Figure 1-10a. *3x3 matrix of feature maps*

0	1	0
0	1	0
0	1	0

Figure 1-10b. *3x3 kernel*

Figure 1-10a is the image or feature map and Figure 1-10b is the kernel. The kernel is the feature extractor, so it will do a dot product on the feature map, resulting in the value 10. This is the first step of our convolution. The images or feature maps are going to be large and so the kernel might not operate on just a 3x3 matrix, but it will take a stride forward to calculate the next convolution operation. Let's look at an extrapolated example of this idea.

2	3	4	2	3
2	3	2	4	1
2	2	3	3	3
7	2	1	8	9
8	8	9	6	5

0	1	0
0	1	0
0	1	0

8	9	9
7	6	15
12	13	17

Figure 1-11. *Feature map, kernel, and resulting output*

As shown in Figure 1-11, a 5x5 feature map was convolved by a 3x3 kernel and that resulted in a 3x3 feature map. That map will again be convolved or converted to some features for a downstream task.

The convolution process also contains a concept of *stride*, which is a hyperparameter telling the kernel how to move around on the feature maps. Given our convolutional neural network, we have a stride of 1. A stride greater than 1 can cause a checkerboard issue in the feature map, with a few pixels getting more attention than the others. We might want or not want this effect, based on our business requirements. A higher stride value can also be used to lower the feature map dimensions.

There is an inherent problem with this convolution. The dimension keeps on shrinking when the convolutions happen. This can be desirable in some sense or some particular use cases but in a few use cases, we might want to retain the original dimensions. We can use the concept of padding on the image or the subsequent feature maps to avoid the issue of dimensionality reduction. Padding is another hyperparameter. We add layers around the images or feature maps.

Figure 1-12b shows how the padding essentially increases the space as well as allows the edge pixel values to be better noticed by the kernels. The convolution process will go through the edge values more than once, thus the information is being carried forward to the next feature maps with more effect. In the case of edges having pixel values, the convolution from kernels will take that value only once, whatever the stride values are on that occasion.

0	3	1
0	3	0
0	4	0

Figure 1-12a. *An edge detector*

0	0	0	0	0
0	4	1	4	0
0	5	1	2	0
0	5	1	1	0
0	0	0	0	0

Figure 1-12b. *A zero-padded feature map*

It is quite interesting how simple padding can change the identification of edges by a kernel. If we assume there is an important edge near the corner of the feature map and don't pad it, the edges won't be detected. This is because, to detect a line or an edge, the kernel or the feature extractor needs to have a similar pattern. In the case of the kernel in Figure 1-12a, which is an edge detector, it needs to find a proper gradient to detect an actual edge. It is because of the gradient from 0 to 4, 0 to 5, and 0 to 5, now it detects the edge. If the padding wasn't there, this gradient wouldn't be there and the kernel would have missed out on an important part.

Receptive Field

When we were working on the concepts of convolutions, there was a feature map and kernel stride. The kernel was extracting features in the space such that the model could interpret the information to its ease. Now we'll consider with an example an 56x56 image with one channel. (This is also written as 56x56x1.) If we try to convert the entire image to features, the entire span of pixels needs to be read. Let's look at a graphic example to sort out this concept.

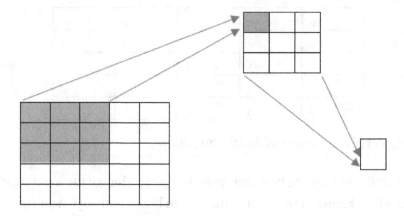

Figure 1-13. *Diagram showing convolution operation*

Figure 1-13 shows a normal 5x5 feature map being convolved by another 3x3 kernel, resulting in a 3x3 kernel, assuming no padding and the stride to be equal to 1. The information carried forward in the step is being carried from the first block of the feature map to the next block of the feature map. That means only the left-most highlighted corner pixel information in Figure 1-13 is contained in the left-most pixel value shown in the 3x3 feature map. This poses a problem because, if we have only one layer and we have a 3x3 kernel, essentially one pixel will have a receptive field of 3x3 in that layer. Interestingly, in the layer after that, if we are using a 3x3 kernel again with stride 1 and zero paddings, the receptive layer is again 3x3 for Figure 1-13, but the 3x3 has already seen the whole image. Thus, it carries all the information from the original image. This concept is called the local and global receptive field.

Local Receptive Field

It is the layer-wise information strength that has been carried forward by the convolutions by the kernel for that step. In the example shown in Figure 1-13, we have a 3x3 kernel, or nine pixels. It will depend on the immediate step and not the entirety.

Global Receptive Field

This is the cumulative information being carried to the last layer of the model. Generally, we want to average out the information from the image to just one value at the end for the sake of simplicity. If we have to predict cats or dogs, it will be easy if we just get one value for the same, rather than a matrix at the end. The last value (1x1xn) or pixel should essentially see all the pixels of the image at the start to be able to perfectly decipher the information.

The example in Figure 1-13 is being convolved by 3x3 and again by 3x3 to reach the final 1x1. Now the 1x1 would have seen the 3x3 figure and that in turn would have seen the 5x5 figure. It's passing the baton and the kernel is helping. The global receptive field here is 5x5.

Pooling

An initial advantage of the convolutional neural network is that it works with parallelism. A fully connected layer would have to worry about an input vector of 25 dimensions even if the image size would be as small as (5x5x1). Even though CNN solves this issue by working on spatial fields, the high dimensionality still can cause the CNN architecture to have a huge number of parameters. Pooling tries to solve this issue by helping with a dimensionality reduction technique and filtering information.

There is an inherent problem with just applying convolution layers in the CNN. The spatial features are captured by the kernel, but a small change in the input feature map will have a large impact on the output feature map. To avoid this challenge, we can use pooling. Depending on the downstream tasks we are doing, we can use max pooling, average pooling, or global average pooling.

Pooling can be thought of as something similar to how the convolution layer applies to the feature map. However, instead of convolution, pooling will work on calculating an average or max of all the values in the region. It can be thought of as a function. There are no parameters that are learned in the pooling section. It is just plain and simple dimensionality reduction in space. It is advisable to use a 2x2 pooling on a higher dimensional feature map, which needs to be greater than 10x10 at least. The reason being the information concentration at this level of a CNN will be too high and a rapid reduction of it by pooling can result in a huge loss.

Max Pooling

Feature maps will contain the information from the image distributed spatially. If we consider a downstream task that needs only the edges, we can try to maximize the edges such that the subsequent feature maps will have a piece of filtered information to focus on. When max pooling has been selected, the prominent features are filtered and passed on to the next layer. When the stride is looking into the stretch of pixels, it selects the one with the highest value, thus the word max.

12	2	4	3
3	2	4	5
5	3	2	3
9	1	2	3

12	5
9	9

Figure 1-14. *Example of max pooling*

Figure 1-14 shows a feature map on the left; we are trying to get a max pool with a stride of 2. The resultant feature map is reduced from 4x4 to 2x2, with just having the prominent feature passing over. In a way, if there are some edges or gradients in an image or feature map, that will take precedence over anything else.

In the case of tasks like classification, we generally follow the process of max pooling, because we need the important edges and gradients to follow and no other irrelevant features messing up the architecture. It is interesting to ponder the fact that color doesn't play a role in most of the classification tasks. For example, a cat can be any color and the model has to understand it is a cat without considering the color.

Average Pooling

We have already established the basic concept of pooling. It gives us a filtration process without increasing any number of learnable parameters. For an average pooling of 2x2 and a stride of 2x2, the strides taking care of a 2x2 block at a time will average the entire section and calculate a mean on top of it. The mean is passed onto the next section of the feature map.

For classification tasks, average pooling is generally not suggested. It can however be used when the images are darker and you want to extract black-to-white transitions.

12	2	4	3
3	3	4	5
5	3	2	5
9	3	2	3

5	4
5	3

Figure 1-15. *Example of average pooling*

Figure 1-15 shows a feature map being pooled by a 2x2 block with a stride of 2. The average of the pooled area is reflected with each 2x2 block from the feature map shown on the left, mapped to one pixel on the right side.

Global Average Pooling

Global average pooling is sometimes used toward the end of the CNN architecture to summarize the feature map to one value. Assume, if you were left with a 5x5 feature map with some depth (channels in the z-direction, assuming x and y are height and width of the feature maps), you could flatten those values out and use a fully connected network to map those like features to get a model. This is a viable option, but taking 5x5, we are forwarding 25 features to a fully connected network and this will take up huge parameters. Instead, you can try to use a global average pooling layer after this step and make it to 1x1, essentially managing all the important features and having a combination. The model parameters we saved by not using a fully connected network can be used to add convolutional layers. This can result in better accuracies.

12	2	4	3
3	5	4	7
5	5	4	5
11	5	2	3

Figure 1-16a. *Sample feature map*

Figure 1-16b. *Result after GAP*

Figure 1-16a shows a feature map with 4x4 dimensions. After pooling, we are left with one single value, as shown in Figure 1-16b. In all cases, we will have a depth, which will have values based on corresponding pooling and feature maps.

Calculation: Feature Map and Receptive Fields

The calculation serves as an important aspect of this coursework. We will be using it to formulate our models and various experimentations. The dimensions of the output feature maps depend on multiple factors, such as stride, kernel size, pooling, padding, input, and output. Let's dive into the details.

Kernel

The kernel is the feature extractor from the feature map or the image. It is initialized in the first forward propagation but the weights are expected to learn and change from the backpropagation to enable it to be a better feature extractor. Once the training sets in, the weights move toward the higher end. This can mean the features it's extracting are important to the cost function and thus the weights. Let's denote the dimension of the kernel with K.

Stride

The step by which the kernel moves around the feature map can be called a stride. Let's denote its value by S.

Pooling

The block in a conventional convolutional neural network architecture tries not to convolve, but aims to capture some kind of information from the spatial feature maps and rapidly reduce the dimension of the feature map for the next step. Let's denote that by mp.

Padding

Padding allows us to keep the dimensions constant after the convolution process has taken place. Let's denote that with P.

Input and Output

Let's denote the input and output as is, where input refers to the feature map at the first step and output refers to the feature map generated.

The first step of getting an architecture started for an image will be to access the size of the image and determine how deep we want to build the network such that the receptive field of the last layer or the output will be equal to the size of the image. In other words, it should have seen what is there in the image to answer anything about the image. If we have an image of size 56x56x3 the kernel to be used can be 3x3x**3**x16. This means the 3x3 kernel has three different sets of initializations and matches up with the channels in the input or preceding layer. This is because it has to know the possible way to mix the features that need to be extracted and used in the network. The formula becomes:

$$HxWxC > KxKxCxC_{next}$$

Here H, W, C are the height, width, and channel of the input image/ feature map.

K is the kernel size and C_{next} is the number of channels in the next step or the batch size. When using CUDA scores to train the CNN, we need to push for the kernel size to be equal to some number defined by 2^n. For example, if we are using 17 as the number of kernels, it will still end up using 32 instead of the closest 2^n number, i.e., 16.

Given all the values and concepts, let's calculate the feature map outputs:

$$Output = \frac{\left[Input + 2P - K\right]}{S} + 1$$

Given, Input = 12x12,
P = 0, K = 3x3, S = 1

$$Output = \frac{[12 + 2*0 - 3]}{1} + 1$$

Output = 10 or 10x10

Calculation of Receptive Field

The receptive calculation at the nth layer will be given by a consolidated formula, as follows

$$Receptive\ Field = \sum_{i=1}^{n}\left((K_i - 1)\prod_{j=0}^{i-1}S_j \right) + 1$$

Let's say we are calculating the receptive field at the second layer So K_1, $K_2 = 3$.

The stride of the kernel is 1 for both occasions and the receptive field comes to 5.

Understanding the CNN Architecture Type

Understanding Types of Architecture

AlexNet

ILSVRC is a competition that was dedicated to computer vision research with its famous ImageNet dataset being used to evaluate models and research in the field. One of the first winners was AlexNet, which had great accuracies in 2010 and 2012. The paper released corresponding to this network revealed usage of the convolutional network, with basic building blocks.

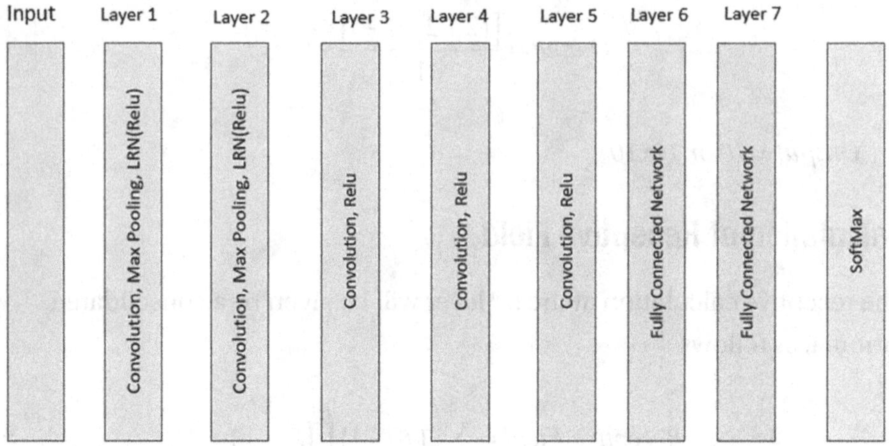

Figure 1-17. *AlexNet architecture*

Figure 1-17. depicts the overview architecture used in AlexNet. The image size taken was 224x224x3 and brought down by a 11x11x3x96 convolution with a stride of 4 to 55x55x96. The max pooling with 3x3 filters and a stride of 2 was used. The first two convolutional layers used LRN and pooling. Following that, the next three layers only used the convolutional and activation layer, followed by two blocks of the fully connected network. The scores were then passed onto the softmax to classify the 1,000 classes.

The most interesting invention in the model architecture is LRN (Local Response Normalization). At the time, sigmoid and tanh were generally used as activation functions, but AlexNet used ReLU. Sigmoid and tanh suffer from saturation at extreme values and the data always has to be centered and normalized to get any gradient pass on during backpropagation. The local response normalization can be thought of as a brightness normalizer and is followed by the ReLU activation.

AlexNet tries to use multiple GPUs to parallelize the training process with increased accuracy. There are two parts to the network that are parallel to each other crossing over at certain sections.

VGG

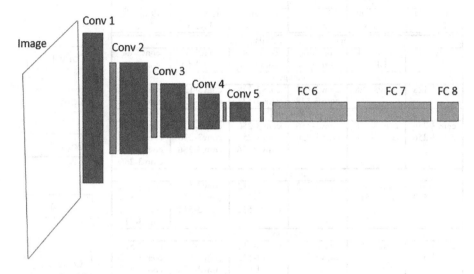

Figure 1-18. *VGG architecture*

Figure 1-18 shows conv 1, conv 2, conv 3, conv 4, and conv 5—a convolution block with ReLU activation. Each block is followed by a max pooling followed by FC 6, FC 7, and FC 8—a fully connected network.

ConvNet Configuration						
A	A-LRN	B	C	D	E	
11 weight layers	11 weight layers	13 weight layers	16 weight layers	16 weight layers	19 weight layers	
input (224 × 224 RGB image)						
conv3-64	conv3-64	conv3-64	conv3-64	conv3-64	conv3-64	Conv 1
	LRN	**conv3-64**	conv3-64	conv3-64	conv3-64	
maxpool						
conv3-128	conv3-128	conv3-128	conv3-128	conv3-128	conv3-128	Conv 2
		conv3-128	conv3-128	conv3-128	conv3-128	
maxpool						
conv3-256	conv3-256	conv3-256	conv3-256	conv3-256	conv3-256	Conv 3
conv3-256	conv3-256	conv3-256	conv3-256	conv3-256	conv3-256	
			conv1-256	**conv3-256**	conv3-256	
					conv3-256	
maxpool						
conv3-512	conv3-512	conv3-512	conv3-512	conv3-512	conv3-512	Conv 4
conv3-512	conv3-512	conv3-512	conv3-512	conv3-512	conv3-512	
			conv1-512	**conv3-512**	conv3-512	
					conv3-512	
maxpool						
conv3-512	conv3-512	conv3-512	conv3-512	conv3-512	conv3-512	Conv 5
conv3-512	conv3-512	conv3-512	conv3-512	conv3-512	conv3-512	
			conv1-512	**conv3-512**	conv3-512	
					conv3-512	
maxpool						
FC-4096						
FC-4096						
FC-1000						
soft-max						

Table 2: **Number of parameters** (in millions).

Network	A,A-LRN	B	C	D	E
Number of parameters	133	133	134	138	144

Figure 1-19. *Layer-wise depiction of a VGG stack*

The architecture started a long-drawn application of going deep with a basic development style. At the time this architecture was proposed, batch normalization was not used so the network suffered from internal covariate shift and gradient lost midway. The architecture uses a combination of 3x3 and followed by 1x1. The reason that stacking three 3x3 is better than one 7x7 is in terms of parameters; if we have C channels, then 3x3 has $27C^2$, but in the case of 7x7 we will end up having $49C^2$.

However, the receptive field ends up with three stacks of 3x3, which is equal to one 7x7, all with more adaptive functions convolving over the feature maps. See Figure 1-19.

The model architecture follows a convolution block-wise architecture that uses a well-defined precedence, used in modern architectures. We have five convolution blocks, and when the feature maps have reduced size, this keeps the information loss minimal. The number of channels increases at each block. Each convolution block is followed by max pooling to reduce the feature maps' dimensions. There is also an addition of 1x1 convolution, which acts as a mixer of all the features of the feature map from the preceding level. This convolution acts as a z-axis dimensionality reduction technique. While continuous usage of 3x3 or 5x5 will increase the number of channels for a feature map, at some point, it needs to get reduced, with proper technique keeping the information loss minimal. For all purposes, 1x1 is used as a cross-channel pooling. It can also be used to increase the number of channels but it is not done in practice. Since it is element-wise multiplication on the feature maps it can easily summarize the content.

VGG was a big improvement over the conventional CNN architecture and was the state-of-the-art algorithm published in 2015.

ResNet

ResNet is an advanced architecture for image downstream tasks and made its a grand entry into the world of CV by winning the ILSVRC competition in 2015. It is also the backbone for a few of the state-of-the-art object detection algorithms like YOLO and faster RCNNs. The paper was put forward by a team from Microsoft. They were able to train deeper with the help of the residual learning framework.

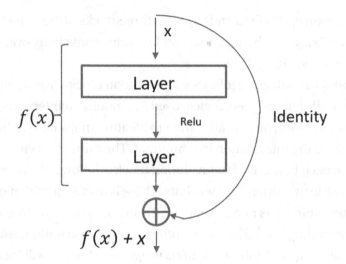

Figure 1-20. *Residual structures*

Figure 1-20 shows the basic structure of the residual framework. The output from the previous layer can be considered as X; it is being passed to the residual function along with an identity function. So, if the residual function is given by the function f(x), we can call the result from both f(x) + x.

The issue that the structure was trying to solve was a degradation of accuracies. The experiments revealed the deeper networks were having their accuracies saturated after the addition of layers and finally falling off. These drops were not from overfitting but due to the problem itself being hard to optimize. In deep neural networks, a common linear problem can be hard to train. For example, if want to build a model with just a linear addition of two numbers and we want the resultant to be the sum, a non-linear counterpart is often easy to optimize, like just adding exponentials to the numbers and the resultants.

Figure 1-21. *ResNet basic architecture*

layer name	output size	18-layer	34-layer	50-layer	101-layer	152-layer
conv1	112×112	7×7, 64, stride 2				
		3×3 max pool, stride 2				
conv2_x	56×56	$\begin{bmatrix} 3\times3, 64 \\ 3\times3, 64 \end{bmatrix} \times 2$	$\begin{bmatrix} 3\times3, 64 \\ 3\times3, 64 \end{bmatrix} \times 3$	$\begin{bmatrix} 1\times1, 64 \\ 3\times3, 64 \\ 1\times1, 256 \end{bmatrix} \times 3$	$\begin{bmatrix} 1\times1, 64 \\ 3\times3, 64 \\ 1\times1, 256 \end{bmatrix} \times 3$	$\begin{bmatrix} 1\times1, 64 \\ 3\times3, 64 \\ 1\times1, 256 \end{bmatrix} \times 3$
conv3_x	28×28	$\begin{bmatrix} 3\times3, 128 \\ 3\times3, 128 \end{bmatrix} \times 2$	$\begin{bmatrix} 3\times3, 128 \\ 3\times3, 128 \end{bmatrix} \times 4$	$\begin{bmatrix} 1\times1, 128 \\ 3\times3, 128 \\ 1\times1, 512 \end{bmatrix} \times 4$	$\begin{bmatrix} 1\times1, 128 \\ 3\times3, 128 \\ 1\times1, 512 \end{bmatrix} \times 4$	$\begin{bmatrix} 1\times1, 128 \\ 3\times3, 128 \\ 1\times1, 512 \end{bmatrix} \times 8$
conv4_x	14×14	$\begin{bmatrix} 3\times3, 256 \\ 3\times3, 256 \end{bmatrix} \times 2$	$\begin{bmatrix} 3\times3, 256 \\ 3\times3, 256 \end{bmatrix} \times 6$	$\begin{bmatrix} 1\times1, 256 \\ 3\times3, 256 \\ 1\times1, 1024 \end{bmatrix} \times 6$	$\begin{bmatrix} 1\times1, 256 \\ 3\times3, 256 \\ 1\times1, 1024 \end{bmatrix} \times 23$	$\begin{bmatrix} 1\times1, 256 \\ 3\times3, 256 \\ 1\times1, 1024 \end{bmatrix} \times 36$
conv5_x	7×7	$\begin{bmatrix} 3\times3, 512 \\ 3\times3, 512 \end{bmatrix} \times 2$	$\begin{bmatrix} 3\times3, 512 \\ 3\times3, 512 \end{bmatrix} \times 3$	$\begin{bmatrix} 1\times1, 512 \\ 3\times3, 512 \\ 1\times1, 2048 \end{bmatrix} \times 3$	$\begin{bmatrix} 1\times1, 512 \\ 3\times3, 512 \\ 1\times1, 2048 \end{bmatrix} \times 3$	$\begin{bmatrix} 1\times1, 512 \\ 3\times3, 512 \\ 1\times1, 2048 \end{bmatrix} \times 3$
	1×1	average pool, 1000-d fc, softmax				
FLOPs		1.8×10^9	3.6×10^9	3.8×10^9	7.6×10^9	11.3×10^9

Figure 1-22. *ResNet layer-wise convolution information*

The identity layers that result from skipping the residual networks provide a second set of available local receptive fields in addition to what we are getting from the residual network. Often termed as the skip connections or the highway networks, they provide a copy of the image to the deep nets. The output from the skip connection and the residual need to have the same dimensions. Thus there is a projection that happens on the output to match the dimensions before being added to the residual functions. Figure 1-21 depicts a ResNet-34 architecture, which can be referred from the table shown in Figure 1-22. The architecture uses a stack of 3,4,6,3, a stack of 64,128,256, and 512 network blocks. There are two types of skip connections—one with solid lines and the other with dotted lines. In most cases if the input dimensions match the output of the residuals, the input is added. This is shown by the solid lines. In the case of dotted lines, the dimensions, if not matched, are controlled by padding in addition to the 1x1 with stride 2. Either way, no parameters are being added for the model to learn.

Another important point to note is that this model uses a stochastic gradient descent with momentum as an optimizer. SGD with momentum is another proven method used in a lot of other state-of-the-art models.

Inception Architectures

Inception has a few versions delivered since 2014. GoogLeNet (often linked with InceptionV1) was the winner of ILSVRC. There are multiple iterations; let's look at inception architecture.

To increase effective local representation in the convolution block, this solution proposed to factorize convolutions. Effectively it suggested using 3x3 after 1x1, reducing the intensity of the activations and reducing correlations. They used a way of factorizing heavier convolution kernels to smaller counterparts trying to achieve the same result. For example, a 5x5 requires $25C^2$ parameters but it can be broken into two sets of 3x3 and still require $18C^2$ parameters. See Figure 1-23. According to the research, 5x5 might have better expressiveness in terms of getting hold of more pixel points in one convolution. Thus, it has a better bird's eye view compared to 3x3. However, the architecture claims that while building a computer vision model, we need to care for translational invariance and a smaller kernel size will help us with that.

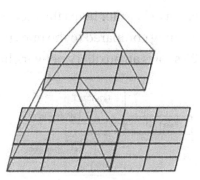

Figure 1-23. *Bird's eye view of convolution*

The factorization is further broken into asymmetric convolutions. The 3x3 can still be broken into 3x1 and then 1x3 convolutions, resulting in at least 9/6 savings. This can be used in a higher set of convolutions as well and is shown generically in Figure 1-24.

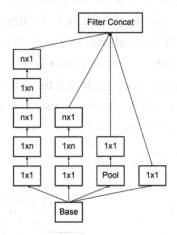

Figure 1-24. *Asymmetric convolutions*

Another important addition to the model architecture were the auxiliary classifiers. They were designed to work on the vanishing gradient issue for deep neural networks. This portion of the network acts as a regularizer to the model, in addition to the batch normalization and dropouts used. The auxiliary classifier helps the model enhance accuracy toward the end of the training compared to the ones that didn't have such branch outs. Figure 1-25 shows an auxiliary network branching out.

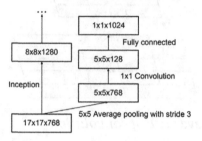

Figure 1-25. *An auxiliary network*

When pooling is applied to an information grid with channels, it is suggested that we increase the number of feature maps according to the reduction we are looking for. Since the information is spatially distributed, we might want to fix the volume of information rather than cutting it off. Imagining a semi-solid cuboid will help here. If we have to reduce the face, we have to elongate in one direction. In our case, it will be the depth. Figure 1-26 shows the grid-reduction technique used in inception architecture. The parallel architecture envisions using pooling and convolutions in parallel and concatenating them.

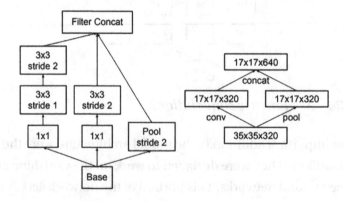

Figure 1-26. *Inception architecture snapshot*

Inception architecture first used stochastic gradient decent with momentum but later used RMSProp to train the model, which achieved better accuracies.

Working with Deep Learning Model Techniques

So far, we have covered the widely used CNN-based models. There are various ways to model architectures for all-purpose computer vision tasks. Let's now discuss a few important concepts that will be useful in the next chapters when we explore multiple problems hands-on.

Batch Normalization

There is an inherent problem that pops up when training a deep neural network. During the training process, the input distribution of the layers is affected to an extent of adopting a different distribution because of the layers from which the inputs are flowing. The gradients affect the weights and those, in turn, have cascading effects on the distributions. This phenomenon is called an *internal covariate shift*. For training deep neural networks using SGD, mini-batch gradient descent is often used. This process reduces the computational requirements at one level and given the mini-batch a representation from the original input, the level of training is good.

Batch normalization is used generally after convolution operations. It can be used after or before a ReLU activation layer, but it is never used after the last layer of the architecture. It enhances the features extracted after each layer. Let's look at an example. Consider an arbitrary layer that has 64 channels and 16 images in one batch. In this scenario, we will be using 16 images to normalize each channel in 64 to get our output.

In the first step of batch normalization, we are trying to shift the data by the mean of the batch and then scale it by the standard deviation of the batch, which corresponds to the channels in the batch. In the second step, we are multiplying the resultant by γ and another parameter β is added to it. These parameters help the neural network determine whether the batch normalization is required and to what extent such features are significant in the next layers. This helps with the flow of gradients and eventually with training.

$$\hat{x} = \frac{x - \mu}{\sigma}$$

$$Output_{BN} = \hat{x} * \gamma + \beta$$

x – Original input

μ - Mean of mini-batch – non-trainable parameters

σ – Standard deviation from mini-batch – non-trainable parameters

γ, β – Parameters normalizing the effects – trainable parameters

Thus, batch normalization incorporates additional trainable parameters into the model architecture.

Ghost batch normalization is a recent development related to batch normalization. This concept is legitimate when we are training in a single GPU and the single GPU has one batch filled in. There can be an issue with the training if someone is using multiple GPUs. Ghost batch normalization helps in the scenario. It will use the GPU as a pivot and will calculate a sample mean and standard deviation to calculate the normalization. This regularizes the data as well, given the fact it is working on a smaller sample size. The randomness in the distribution of images due to the use of multiple GPUs will change the loss function to an extent. Batch normalization has proven to be a great mechanism for models to go deeper; it provides manifold benefits to modern-day architectures.

Dropouts

Dropouts are generally used for one-dimensional networks. In the case of convolutional neural networks, a few models do have dropouts and they can be used after every convolutional layer except the last few layers. The last few layers have smaller dimensions and the image will have been represented. It can act as a regularizer.

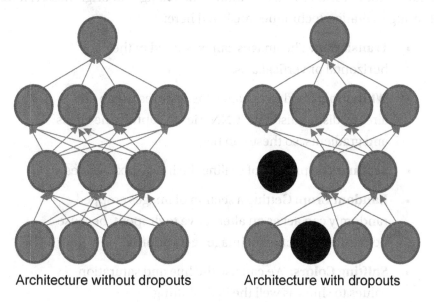

Architecture without dropouts   Architecture with dropouts

Figure 1-27. *Dropout examples in a fully connected network*

Figure 1-27 shows the dropout application. On the left side of the image is a general neural network. The right side shows dropouts with all the layers except the last one.

Data Augmentation Techniques

Deep learning architectures are best with volumes of data in order to generalize on the cost function. In real-world scenarios, sufficient data is often not available. Data augmentation is used to generate data points without affecting the actual labels or basic distinguishable features of the image. For example, if we want to augment a picture of a cat, we must provide the classifier with not just the image of a tail. The tail can be a distinguishable feature within the same species, but it can belong to other objects as well.

The training data needs to be representative of the population. Images can have a lot of variations, translations, and changes in brightness. A few of the augmentation techniques are listed here:

- **Translation:** The images can be shifted in the horizontal or vertical axis.

- **Whitening:** Used to enhance the features which are already distinguishable. CNN tries to capture the edges and gradients, so these can help.

- **Scaling:** The process of scaling the image pixel values.

- **Random Crop:** Getting a section of images cropped randomly can act as an alternative to dropout, because no particular section of images gets priority.

- **Shifting Colors:** We can use the hue and saturation values to shifts as well the RGB shifting.

- **Elastic distortions:** We can try to compute displacement interpolations, shifting a few points in the vertical and horizontal axes.

- **Mixup:** In some model scenarios of classification, we can use this technique to create a blend of two classes so that the CNN can come up with a linear differentiating ability.

- **Patch Gaussian:** This technique adds random noise to selected patches in an image. This enhances the robustness of the model.

- **Reinforcement Based Augmentation:** We can create the model and decide on policies for augmentations. This is an expensive process and should be used with caution.

There are multiple data augmentation techniques, and most of them are modulated by the kind of robustness we want and that the test data is expecting. It is an important aspect of any deep learning-based modeling.

Introduction to PyTorch

Deep learning modeling needs high computations and properly defined frameworks for optimization. There are a few existing frameworks that are quite popular with the research and the developer community. In all the frameworks, they have made sure the useful and recurring functions are readily available. A framework essentially sets up the packages and functions that will help developers create end-to-end deep learning models with ease.

PyTorch is a framework developed in C++ and Python, so it is optimized and faster. It uses tensors for most of its development. It also aids in manipulating tensors in different processor architectures. It supports parallel processing and processing in CUDA cores, which speeds up the training of computer vision models.

The following sections discuss how to use PyTorch.

Installation

Visit pytorch.org and select the system for which you want to install PyTorch. For example, if you want to install the package via conda or pip, it's best to have CUDA installed and to leverage that.

```
<conda> / <pip> install pytorch/torch <configuration>
```

Basic Start

```
import numpy as np
import torch
print(torch.__version__)
>> 1.9.0+cu102
```

Here are the basic operations around converting from NumPy to a tensor.

```
x1 = [[1,1],[2,2]]
print("Type of data :{}".format(type(x1)))
print(x1)
x1 = np.array(x1)
print("Type of data :{}".format(type(x1)))
print(x1)
```

```
Type of data :<class 'list'>
[[1, 1], [2, 2]]
Type of data :<class 'numpy.ndarray'>
[[1 1]
 [2 2]]
```

Figure 1-28a. *Converting NumPy to a tensor*

Here, we are creating a list and converting that to a NumPy array.

```
x_tensor = torch.tensor(x1)
print("Type of data :{}".format(type(x_tensor)))
print(x_tensor)
```

```
Type of data :<class 'torch.Tensor'>
tensor([[1, 1],
        [2, 2]])
```

Figure 1-28b. *Converting a list to a tensor*

Here, we are converting the NumPy data to a tensor using `torch.tensor`.

```
x1 =np.array([[2,2],[2,2]])
x_tensor = torch.Tensor(x1)
print("Type of data :{}".format(type(x_tensor)))
print(x_tensor)
```

```
Type of data :<class 'torch.Tensor'>
tensor([[2., 2.],
        [2., 2.]])
```

Figure 1-28c. *Example of a tensor generated from an array*

Next, we are converting the Numpy array to a Torch tensor using `torch.Tensor`, which is a constructor that converts existing data to a tensor or creates an uninitialized data tensor (`torch.empty`).

```
x1 =np.array([[1,2],[1,2]])
x_tensor = torch.from_numpy(x1)
print("Type of data :{}".format(type(x_tensor)))
print(x_tensor)
```

```
Type of data :<class 'torch.Tensor'>
tensor([[1, 2],
        [1, 2]])
```

Figure 1-28d. *Example of a tensor generated from an array*

This function is also used to create a tensor from a NumPy array. Now let's look at some functions in the matrix multiplications:

```
zero_t = torch.zeros((2,2))
print(zero_t)
one_t = torch.ones((2,2))
print(one_t)
rand_t = torch.rand(2,2)
print(rand_t)
```

```
            tensor([[0., 0.],
                    [0., 0.]])
            tensor([[1., 1.],
                    [1., 1.]])
            tensor([[0.1928, 0.3161],
                    [0.2537, 0.1133]])
```

Figure 1-28e. *Examples of a few tensor generators*

We are creating a Torch tensor of all zeros, all ones, and another random value tensor. We can use that for basic linear algebra operations.

```
print(one_t + rand_t)
```

```
            tensor([[1.1928, 1.3161],
                    [1.2537, 1.1133]])
```

Figure 1-28f. *Example of tensor addition*

```
print(one_t*rand_t)
```

```
            tensor([[0.1928, 0.3161],
                    [0.2537, 0.1133]])
```

Figure 1-28g. *Example of tensor multiplication*

The results show the addition of two matrices and multiplication, looking at the matrix operations in detail.

```
array1 = torch.tensor([1,2,4])
array2 = torch.tensor([2,3,4])
print(torch.dot(array1,array2))
```

```
>> tensor(24)
```

The torch.dot function computes the dot product of two one-dimensional tensors.

```
array1 = torch.tensor([1,2,4]).reshape(1,-1)
print(array1.shape)
array2 = torch.tensor([2,3,4]).reshape(-1,1)
print(array2.shape)
print(torch.matmul(array2,array1))
```

```
        torch.Size([1, 3])
        torch.Size([3, 1])
        tensor([[ 2,  4,  8],
                [ 3,  6, 12],
                [ 4,  8, 16]])
```

Figure 1-28h. *Example of tensor dot products*

Summary

This chapter has explained the basic ways of doing a dot product and matrix multiplication. We can consider the channels and inputs to be three-dimensional tensors. The modeling framework uses dot products, element-wise multiplications, and other linear algebraic operations in computer vision models. We will be taking a deeper dive into modeling with Torch in the following chapters.

We went through the basics of convolutional neural networks and how they help in understanding images.

With this, we conclude our discussion of the computer vision concepts. We start with the applications and projects in the next chapter.

CHAPTER 2

Image Classification

The last chapter discussed several important concepts in computer vision. A few of the best practices in the field of computer vision were discussed as well, so it is time to put them into action. This chapter sets the tone for multiple applications in the field of computer vision. We start with a basic explanation of how to start using the Torch components to build a model, define a loss function, and train.

An object that needs to be identified by its name involves the process of classification. We have encountered problems in all facets of data science that involve a classification requirement. It can be as simple as classifying an image on your phone as to whether it is a picture of mountains or sea, or whether it's a bird or a dog. Classification is one of the most basic yet most powerful concepts. Let's look at how classification is set up by a computer vision model:

1. Detecting edges

2. Detecting gradients

3. Identifying textures

4. Identifying patterns

5. Forming parts of objects

A model needs to associate a name with a particular object in an image. It does this by following a structured knowledge extraction mechanism and then by regenerating its input for the decision-making process.

© Akshay Kulkarni, Adarsha Shivananda, and Nitin Ranjan Sharma 2022
A. Kulkarni et al., *Computer Vision Projects with PyTorch*,
https://doi.org/10.1007/978-1-4842-8273-1_2

Topics to Cover

1. Data preparation methods

2. Data augmentation techniques

3. Using batch normalization and dropouts

4. Comparing activation functions

5. Setting up models and their variations

6. Training process

7. Running inference and comparing model results

Defining the Problem

We will be checking x-ray images of lungs and classifying them, with the help of computer vision modeling techniques, as either having pneumonia or being normal. Since this is a healthcare problem it is always best to let the model overpredict. We need to predict with the highest form of accuracy and should have nearly 100% recall if possible, with a high precision score as well. We need to be sure to diagnose any possible case of infection and not misclassify an infected lung as healthy because of small margins. Softmax logits can often be used to determine the prediction rather than a softmax function deciding the class. This is a critical decision based on experience with the data as well as the behavior of the model.

A major problem that clouds these kinds of image classification problems is the availability of properly annotated data. Image classification by convolutional neural networks helps with multiple downstream tasks. A model trained on some data can be used to fine-tune other similar data and be used for prediction purposes. There are multiple open-source image repositories, but for the most industrial purpose they give us a starting point. We have to use our task-specific images as well.

Overview of the Approach

We will use convolutional neural networks to solve the classification problem. We will try to fit in variations of the processes and check for higher accuracy and stable results. The concepts learned in Chapter 1 will be used extensively in this process. This is strictly an experiment and we will only set the baseline standards for the approach that needs to be iterated.

This approach involves the following steps:

1. Downloading data from the source and placing it in the root directory.

2. Checking for data sanity, configurable information such as shape, size, and distribution of the images.

3. Initializing data loader functionalities for training and testing.

4. Defining the model architecture and validating it.

5. Defining the functions for training and testing.

6. Defining the optimizer for training and other training information such as regularizers, epochs, batches, etc.

7. Training and checking for loss and accuracy patterns to understand the stability of the architecture and model training process.

8. Deciding over multiple stages of improvements or changes, which one to select for further tuning or production.

A graphical overview of this approach is shown in Figure 2-1, which can be taken for the solution.

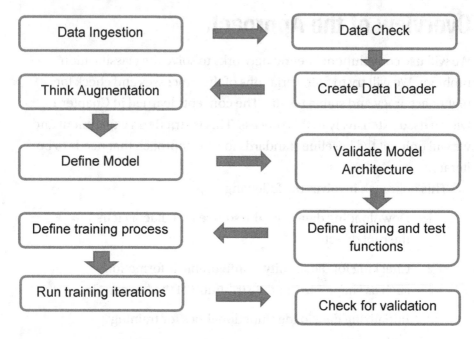

Figure 2-1. *Image classification pipeline*

Creating an Image Classification Pipeline

There can be multiple ways to proceed with a simple classification problem. Since we are working with deep learning models that can work out features from spatial patterns, going deep into the network can help. We also need to look at other strategies like learning rate modulations and regularizing techniques to help the model. Let's look at the strategies that we can apply while considering the complexities of the problem:

1. We can check on the data availability, based on how much data is available and the complexity of the problem. We can then decide if we have to oversample.

2. The validation data and the type of image data we get helps with the data augmentation strategies.

3. We need to check for image sizes and the object within the image size so that we can understand the model architecture better.

4. We need to formulate our strategies around the production infrastructure, on which the model needs to run, as well as the kind of latency we need.

5. The model strategy will also require us to figure out the accuracy, whether a higher recall is required or we need higher precision.

6. The training time and cost in infrastructure also need to be taken into consideration while building the model.

We will try four progressive approaches to deal with the image classification problem at hand. We will increase the complexity of our solution in a step-like manner to see the impact of the individual process. Let's look at the first strategy.

First Basic Model
Data

This use case creates an image classifier over a dataset of x-ray images of lungs with pneumonia and normal lungs. We will be downloading the dataset and placing it in our local directory, which is accessible by the Python compiler. If you are using Google Colab, you can use Google Drive as storage, which can be mounted on Colab.

In this problem set, we use open source data, which can be found at https://www.kaggle.com/tolgadincer/labeled-chest-xray-images.

The dataset is divided into test and train folders, with each further divided into NORMAL and PNEUMONIA categories.

- The number of train samples in the NORMAL category is 1349.

- The number of train samples in the PNEUMONIA category is 3883.

- The number of test samples in the NORMAL category is 234.

- The number of test samples in the PNEUMONIA category is 390.

Let's look at a sample image from the NORMAL folder to check on the quality and positioning of the image. Figure 2-2 shows a random 2283x2120 image from the NORMAL train folder. Since this is generated from mpimg, the color when displayed in Jupyter notebook is different. You can also use another command to display the image, cv2.imshow(), in its place.

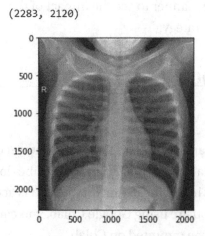

Figure 2-2. *Sample image from the train data of normal lungs*

Let's now look at an example image from the PNEUMONIA train folder. Figure 2-3 shows a 776x1224 image from this class.

(776, 1224)

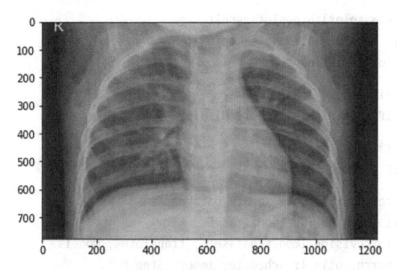

Figure 2-3. *Sample image of infected lungs*

Let's start coding with basic imports. These are required for the entire flow to work. GPU is better for a faster training process but CPU should also work.

We need to install PyTorch, with CUDA support in case we are using local CUDA cores for our training process. We need to be careful about all the objects placed in CUDA for processing and all the objects placed in the CPU for processing. The intermingling of data across different processor types is not supported unless they are specifically cast.

We need to import a few custom libraries for these classification problems. Let's enlist them in order.

```
import os

import numpy as np
import cv2

import matplotlib.pyplot as plt
import matplotlib.image as mpimg
%matplotlib inline

from PIL import Image
from IPython.display import display

import torch
import torch.nn as nn

from torch.utils.data import DataLoader
import torch.nn.functional as F
from torchvision import datasets, transforms, models
from torch.optim.lr_scheduler import StepLR
from torchsummary import summary
from tqdm import tqdm
```

After we have imported all the required libraries, we can start linking the data from the directory. We start by unzipping the files to the folders. If we are using Google Colab for this process, we can use the following commands to mount the Google Drive to Colab and use the data stored there.

```
from google.colab import drive
drive.mount('/content/gdrive')

!unzip <zipped file location>
```

This will bring the data to the Colab location, so the model can easily use it. After this, we set the data path for the data directory irrespective of the system we will be using.

```
data_path = '/content/chest_xray'
```

Data Exploration

We will now explore and check the sanity of the data. We have to assign train and test folders that can be used in the model. In image classification, there is no specific image-wise annotation. We can use the folder name as the class name if the images are segregated by folders. There can be another variation where we can see all images in one folder and then specify which image path belongs to which class.

```python
class_name = ['NORMAL','PNEUMONIA']
def get_list_files(dirName):
    '''

    input - directory location
    output - list the files in the directory
    '''
    files_list = os.listdir(dirName)
    return files_list

files_list_normal_train = get_list_files(data_path+'/
train/'+class_name[0])
files_list_pneu_train = get_list_files(data_path+'/
train/'+class_name[1])

files_list_normal_test = get_list_files(data_path+'/
test/'+class_name[0])
files_list_pneu_test = get_list_files(data_path+'/
test/'+class_name[1])
```

We are hard-coding the class names to be NORMAL and PNEUMONIA since the folders are arranged in that fashion.

```python
print("Number of train samples in Normal category {}".
format(len(files_list_normal_train)))
print("Number of train samples in Pneumonia category {}".
format(len(files_list_pneu_train)))
```

```
print("Number of test samples in Normal category {}".
format(len(files_list_normal_test)))
print("Number of test samples in Pneumonia category {}".
format(len(files_list_pneu_test)))
```

```
Output:
Number of train samples in Normal category 1349
Number of train samples in Pneumonia category 3883
Number of test samples in Normal category 234
Number of test samples in Pneumonia category 390
```

Now that we have extracted the images and located the path, let's see how to view the sample images from NORMAL and PNEUMONIA folders.

```
rand_img_no = np.random.randint(0,len(files_list_normal_train))
img = data_path + '/train/NORMAL/'+ files_list_normal_
train[rand_img_no]
print(plt.imread(img).shape)
#display(Image.open(img,'r'),)

img = mpimg.imread(img)
imgplot = plt.imshow(img)
plt.show()
```

The output here is the image shown in Figure 2-2.

```
img = data_path + '/train/PNEUMONIA/'+ files_list_pneu_
train[np.random.randint(0,len(files_list_pneu_train))]
print(plt.imread(img).shape)

img = mpimg.imread(img)
imgplot = plt.imshow(img)
plt.show()
```

The output in this case is the image is shown in Figure 2-3.

Data Loader

Since we have explored the data, now let's set up the data loaders for training purposes. In this variation, we will not be using augmentation to help with training regularization. We will just resize and crop the images to a uniform size of 224x224. This starting point for the images is not set in stone; you can use a different size if you want.

Apart from the size and cropping of the image, we will also look at converting the images to tensors for the PyTorch framework. We will try to normalize the images with mean and standard deviation values. If we are considering three channels per image, then we need to provide three values for one channel. We need one combination of mean and standard deviation.

Here's the code:

```
train_transform = transforms.Compose([
    transforms.Resize(224),
    transforms.CenterCrop(224),
    transforms.ToTensor(),
    transforms.Normalize([0.485, 0.456, 0.406],
                         [0.229, 0.224, 0.225])
test_transform = transforms.Compose([
    transforms.Resize(224),
    transforms.CenterCrop(224),
    transforms.ToTensor(),
    transforms.Normalize([0.485, 0.456, 0.406],
                         [0.229, 0.224, 0.225])])

train_data = datasets.ImageFolder(os.path.join(data_path,
'train'), transform= train_transform)
test_data = datasets.ImageFolder(os.path.join(data_path,
'test'), transform= test_transform)
```

```
train_loader = DataLoader(train_data,
                          batch_size= 16, shuffle= True, pin_
                          memory= True)
test_loader = DataLoader(test_data,
                          batch_size= 1, shuffle= False, pin_
                          memory= True)

class_names = train_data.classes

print(class_names)
print(f'Number of train images: {len(train_data)}')
print(f'Number of test images: {len(test_data)}')

Output:
['NORMAL', 'PNEUMONIA']
Training images available: 5232
Testing  images available: 624
```

We are using the default data loader of PyTorch. We will create two sets of data loaders, one for the training dataset and another for the testing set. The batch size is variable in each case, depending on the GPU and RAM of the system. We can shuffle the training data since no particular order is necessary. In the case of the testing data, the shuffle needs to be turned off.

The pin memory argument helps if someone needs to transfer the dataset previously loaded in the CPU to a GPU. The process is faster when pin memory is enabled.

We are using the data loader to transform functionalities in the data and use them in the train function later. The image folder is generally used when the images are arranged according to the name of the class in the folders.

Define the Model

We will define our model architecture with convolution blocks and use ReLU as an activation layer. The baseline model will have 12 convolution blocks, including one convolution block to set up the input and one for the output. The first three convolution blocks have one max pooling function to drop down from the high dimensions of the image to lower dimensions by filtering information.

The model definition follows:

```
class Net(nn.Module):
    def __init__(self):
        super(Net, self).__init__()
        # Input Block
        self.convblock1 = nn.Sequential(
            nn.Conv2d(in_channels=3, out_channels=8, kernel_
            size=(3, 3),
                        padding=0, bias=False),
            nn.ReLU(),
            #nn.BatchNorm2d(4)
        )
        self.pool11 = nn.MaxPool2d(2, 2)

        # CONVOLUTION BLOCK
        self.convblock2 = nn.Sequential(
            nn.Conv2d(in_channels=8, out_channels=16, kernel_
            size=(3, 3),
                        padding=0, bias=False),
            nn.ReLU(),
            #nn.BatchNorm2d(16)
        )
```

```python
    # TRANSITION BLOCK

    self.pool22 = nn.MaxPool2d(2, 2)

    self.convblock3 = nn.Sequential(
        nn.Conv2d(in_channels=16, out_channels=10, kernel_
        size=(1, 1), padding=0, bias=False),
        #nn.BatchNorm2d(10),
        nn.ReLU()
    )
    self.pool33 = nn.MaxPool2d(2, 2)

    # CONVOLUTION BLOCK
    self.convblock4 = nn.Sequential(
        nn.Conv2d(in_channels=10, out_channels=10, kernel_
        size=(3, 3), padding=0, bias=False),
        nn.ReLU(),
        #nn.BatchNorm2d(10)
    )

    self.convblock5 = nn.Sequential(
        nn.Conv2d(in_channels=10, out_channels=32, kernel_
        size=(1, 1), padding=0, bias=False),
        #nn.BatchNorm2d(32),
        nn.ReLU(),

    )

    self.convblock6 = nn.Sequential(
        nn.Conv2d(in_channels=32, out_channels=10, kernel_
        size=(1, 1), padding=0, bias=False),
        nn.ReLU(),
        #nn.BatchNorm2d(10),

    )
```

```python
self.convblock7 = nn.Sequential(
    nn.Conv2d(in_channels=10, out_channels=10, kernel_
    size=(3, 3), padding=0, bias=False),
    nn.ReLU(),
    #nn.BatchNorm2d(10)

)

self.convblock8 = nn.Sequential(
    nn.Conv2d(in_channels=10, out_channels=32, kernel_
    size=(1, 1), padding=0, bias=False),
    #nn.BatchNorm2d(32),
    nn.ReLU()
)
self.convblock9 = nn.Sequential(
    nn.Conv2d(in_channels=32, out_channels=10, kernel_
    size=(1, 1), padding=0, bias=False),
    nn.ReLU(),
    #nn.BatchNorm2d(10),

)


self.convblock10 = nn.Sequential(
    nn.Conv2d(in_channels=10, out_channels=14, kernel_
    size=(3, 3), padding=0, bias=False),
    nn.ReLU(),
    #nn.BatchNorm2d(14),

)

self.convblock11 = nn.Sequential(
    nn.Conv2d(in_channels=14, out_channels=16, kernel_
    size=(3, 3), padding=0, bias=False),
```

```
        nn.ReLU(),
        #nn.BatchNorm2d(16),
    )

    # OUTPUT BLOCK
    self.gap = nn.Sequential(
        nn.AvgPool2d(kernel_size=4)
    )

    self.convblockout = nn.Sequential(
            nn.Conv2d(in_channels=16, out_channels=2, kernel_
            size=(4, 4), padding=0, bias=False),
    )

def forward(self, x):
    x = self.convblock1(x)
    x = self.pool11(x)
    x = self.convblock2(x)
    x = self.pool22(x)
    x = self.convblock3(x)
    x = self.pool33(x)
    x = self.convblock4(x)
    x = self.convblock5(x)
    x = self.convblock6(x)
    x = self.convblock7(x)
    x = self.convblock8(x)
    x = self.convblock9(x)
    x = self.convblock10(x)
    x = self.convblock11(x)
    x = self.gap(x)
    x = self.convblockout(x)

    x = x.view(-1, 2)
    return F.log_softmax(x, dim=-1)
```

In this approach, we are creating a Net class that enables options for multiple inheritances using the Python super functionality. We start with the input convolutional block; the number of input channels is set to 3 and the output channels are set to 8. These arguments can be tweaked, but should be in line with the architecture and core availabilities. We use the kernel with 3x3 convolutions since, as discussed earlier, it is one of the most efficient ways of convolving. In a few blocks, we can also see a 1x1, which helps reduce the feature maps in the z-direction by proposing a combination of all the feature maps.

Here's the model explanation in detail:

1. The input block for the model receives a three-channel 224x224 input and uses a convolution by 3x3 to generate 222x222 and eight channels. This is followed by the ReLU activation layer. We are not using padding for this model architecture.

2. After the input, we call on the max pooling function to reduce the feature map size to 111x111.

3. After the pooling function works on the feature maps, we are convolving the feature maps by 3x3 to generate 16 channels from 8 and reducing the feature map dimension to 109x109.

4. Once we have used a convolution block to get 16 channels, we use the max pooling function again to bring the feature map dimension to 54x54.

5. We then use a transition block (for the first time in the network) to reduce the number of channels from 16 to 10, followed by another max pooling function.

6. Once we have finished using max pooling and the feature map dimension is now 27x27, we use a 3x3 kernel to convolve and create an equal number of feature maps.

7. The fifth and sixth convolution blocks are transition layers, wherein we increase the number of layers from 10 to 32 and back to 10. There is no padding as usual for this.

8. The seventh convolution block is used for 3x3 convolution operations, but the channel size remains the same again.

9. We have similar operations in the eighth and ninth convolution blocks. Using the transition convolutional operations, we are moving the channels from 10 to 32 and back to 10 again.

10. We have added a 3x3 convolution block, which is tenth in line. We increase the number of feature maps from 10 to 14.

11. The penultimate building block of the architecture is using a 3x3 kernel to move the number of channels from 14 to 16.

12. In the output block, we use average pooling to bring in two-four units from 19x19. That can be used for the binary classification. Following our average pooling, we use a convolution block of the same kernel size as of the feature map dimension to bring that into a single unit for the output.

13. Finally, we use the logarithmic `softmax` function to produce the output. It is a scaled output and we use the `argmax` function to determine the class per batch element.

For the purposes of this architectural design, we have kept the addition of bias set to false. This means no bias is added to the calculation for all the neural components formed in the network. We can, however, experiment with bias. In most places, it shouldn't affect the network much, as long as the data is centered and normalized.

Let's look at the signature of the model as output from the summary feature. We can also put the model in a GPU if one is available for processing.

```
use_cuda = torch.cuda.is_available()
device = torch.device("cuda" if use_cuda else "cpu")
print("Available processor {}".format(device))
model = Net().to(device)
summary(model, input_size=(3, 224, 224))
```

Available processor cuda

Layer (type)	Output Shape	Param #
Conv2d-1	[-1, 8, 222, 222]	216
ReLU-2	[-1, 8, 222, 222]	0
MaxPool2d-3	[-1, 8, 111, 111]	0
Conv2d-4	[-1, 16, 109, 109]	1,152
ReLU-5	[-1, 16, 109, 109]	0
MaxPool2d-6	[-1, 16, 54, 54]	0
Conv2d-7	[-1, 10, 54, 54]	160
ReLU-8	[-1, 10, 54, 54]	0
MaxPool2d-9	[-1, 10, 27, 27]	0
Conv2d-10	[-1, 10, 25, 25]	900
ReLU-11	[-1, 10, 25, 25]	0
Conv2d-12	[-1, 32, 25, 25]	320
ReLU-13	[-1, 32, 25, 25]	0
Conv2d-14	[-1, 10, 25, 25]	320

ReLU-15	[-1, 10, 25, 25]	0
Conv2d-16	[-1, 10, 23, 23]	900
ReLU-17	[-1, 10, 23, 23]	0
Conv2d-18	[-1, 32, 23, 23]	320
ReLU-19	[-1, 32, 23, 23]	0
Conv2d-20	[-1, 10, 23, 23]	320
ReLU-21	[-1, 10, 23, 23]	0
Conv2d-22	[-1, 14, 21, 21]	1,260
ReLU-23	[-1, 14, 21, 21]	0
Conv2d-24	[-1, 16, 19, 19]	2,016
ReLU-25	[-1, 16, 19, 19]	0
AvgPool2d-26	[-1, 16, 4, 4]	0
Conv2d-27	[-1, 2, 1, 1]	512

==

Total params: 8,396
Trainable params: 8,396
Non-trainable params: 0

Input size (MB): 0.57
Forward/backward pass size (MB): 11.63
Params size (MB): 0.03
Estimated Total Size (MB): 12.23

This is the summary of the model created from the model design. We are calculating the workflow by giving the input dimensions expected by the model and validating them in the process.

We need to note the model's trainable and non-trainable parameters, which will be a factor for our training as well as the volume of weights going into a production infrastructure. The model size is given as 11.63MB.

The Training Process

After defining the model and the data loader, we have come down to training. The training process will include the following important processes:

1. Initializing the gradient for the model workflow.

2. Getting a prediction from the current model or the forward pass given the current weights of the model. Initially, the weights are randomly assigned from a distribution, using the Xavier or He initialization. (For a ReLU activation network, He is used, whereas for sigmoid, Xavier is used.)

3. Once the forward pass is complete, a loss is calculated that measures how far off the predictions are from the actual values.

4. We then calculate the backward propagation given the accumulated loss.

5. Once the backward propagation loss computations are done, we move toward the optimizer step, which will use the learning rate and other arguments to refresh and update the weights of the model.

Here's the code to prepare the data for training and testing:

```
train_losses = []
test_losses = []
train_acc = []
test_acc = []

def train(model, device, train_loader, optimizer, epoch):
    model.train()
    pbar = tqdm(train_loader)
```

```
    correct = 0
    processed = 0
    for batch_idx, (data, target) in enumerate(pbar):
        # get data
        data, target = data.to(device), target.to(device)

        # Initialization of gradient
        optimizer.zero_grad()
        # In PyTorch, gradient is accumulated over backprop and
        even though thats used in RNN generally not used in CNN
        # or specific requirements
        ## prediction on data
        y_pred = model(data)

        # Calculating loss given the prediction
        loss = F.nll_loss(y_pred, target)
        train_losses.append(loss)

        # Backprop
        loss.backward()
        optimizer.step()
        # get the index of the log-probability corresponding to
        the max value
        pred = y_pred.argmax(dim=1, keepdim=True)
        correct += pred.eq(target.view_as(pred)).sum().item()
        processed += len(data)

        pbar.set_description(desc= f'Loss={loss.item()} Batch_
        id={batch_idx} Accuracy={100*correct/processed:0.2f}')
        train_acc.append(100*correct/processed)

def test(model, device, test_loader):
    model.eval()
    test_loss = 0
```

```
correct = 0
with torch.no_grad():
    for data, target in test_loader:
        data, target = data.to(device), target.to(device)
        output = model(data)
        test_loss += F.nll_loss(output, target,
        reduction='sum').item()
        pred = output.argmax(dim=1, keepdim=True)
        correct += pred.eq(target.view_as(pred)).sum().item()
test_loss /= len(test_loader.dataset)
test_losses.append(test_loss)

print('\nTest set: Average loss: {:.4f}, Accuracy: {}/{}
({:.2f}%)\n'.format(
        test_loss, correct, len(test_loader.dataset),
        100. * correct / len(test_loader.dataset)))

test_acc.append(100. * correct / len(test_loader.dataset))
```

The code block essentially creates two functions that can be used for training purposes and evaluates the model based on how efficiently it works on the test data. The code block also creates two sets of accuracies and two sets of losses from the training and testing data while the training process is going on. This helps determine how the model is behaving and measures its robustness.

Let's go through the code block and decipher the steps:

1. The train function sets up the model for training.

2. We place the data in the GPU if the model is in GPU, or CPU if the model is in CPU. Initializing the device ensures the data and the model are on the same device during training.

3. We set the gradient to 0 every time a new batch comes in, since PyTorch by default tries to accumulate a gradient, and it is not good for convolutional neural networks. The gradient-accumulation process can be leveraged for temporal-based models and architectures.

4. We use the data loader to generate batches of images and pass them on to the model for training.

5. We calculate the prediction from the forward pass and place that in a variable. Once that is done, we calculate the losses due to the prediction from the model.

6. The computed loss helps in backpropagation and helps the optimizer take an update of the weights of the models according to the direction of the steepest ascent/descent.

7. The prediction class is computed from the logarithmic softmax function, by taking the maximum of the index and computing the value accordingly.

8. To compute the test accuracies and the losses, we perform the same process but keep the model under evaluation.

9. We don't update the weights while calculating the loss from the test samples.

Now that we have computed the functions for our loss computation, back propagation, and weights, we can initiate the optimizers and schedulers to begin training. Here's the code for the training process:

```
model = Net().to(device)
optimizer = torch.optim.SGD(model.parameters(), lr=0.01,
momentum=0.9)
scheduler = StepLR(optimizer, step_size=6, gamma=0.5)

EPOCHS = 15
for epoch in range(EPOCHS):
    print("EPOCH:", epoch)
    train(model, device, train_loader, optimizer, epoch)
    scheduler.step()
    print('current Learning Rate: ', optimizer.state_dict()
    ["param_groups"][0]["lr"])
    test(model, device, test_loader)
```

We use a stochastic gradient optimizer with momentum to fit in the model. We are also using the schedular to periodically change the learning rate for the optimizer. This can indirectly help with faster convergence. The number of epochs is also dependent on how we want to train the model, and whether the computation time suits our purposes or not. We will look at the saturation of loss function before stopping the training process.

Look at the output snippet shown in Figure 2-4.

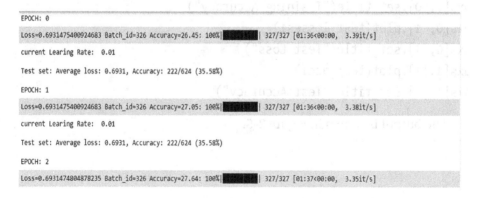

Figure 2-4. Snapshot of how the output will look

The output from the code block will generate the training information such as train and test loss and show the accuracies. We note the highest accuracies and will eventually want to save the model weights at that point only.

This approach might not have given us the best result, but it has established the workflow that we will use to get better accuracies step by step. In this model, we are getting a very low accuracy of 38% in the test dataset. Let's analyze the loss pattern for the testing and training data to figure out the problem.

The code snippet for producing the visualization of losses is as follows:

```
train_losses1 = [float(i.cpu().detach().numpy()) for i in
train_losses]
train_acc1 = [i for i in train_accuracies]
test_losses1 = [i for i in test_losses]
test_acc1 = [i for i in test_accuracies]

fig, axs = plt.subplots(2,2,figsize=(16,10))
axs[0, 0].plot(train_losses1,color='green')
axs[0, 0].set_title("Training Loss")
axs[1, 0].plot(train_acc1,color='green')
axs[1, 0].set_title("Training Accuracy")
axs[0, 1].plot(test_losses1)
axs[0, 1].set_title("Test Loss")
axs[1, 1].plot(test_acc1)
axs[1, 1].set_title("Test Accuracy")
```

The output is shown in Figure 2-5.

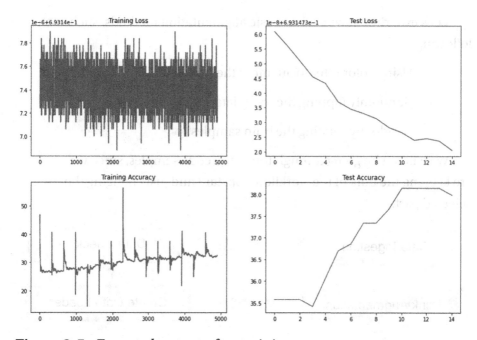

Figure 2-5. *Expected output after training*

From Figure 2-5, we can see that even though the test accuracies have increased over the epochs and losses have consistently decreased, it has not reached the desired state. The training loss shows that the model is highly unstable in real-time. It's time to rethink this approach and build on top of this workflow.

The Second Variation of Model

Let's start by augmenting the data and seeing if there are changes in the accuracies. There are multiple augmentation processes; we should select the one that best meets our business requirements. We need to be careful because too much augmentation can have an impact on the optimization.

Let's experiment with a few basic augmentation methods, such as the following:

- Using color jitter to augment the train data.

- Randomly flipping the train data.

- Randomly rotating the train samples.

We will not be going through the entire code samples, since we are keeping the bulk of the workflow constant and just changing the selected part.

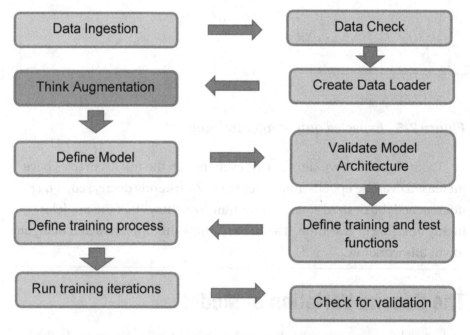

Figure 2-6. *Image classification pipeline updated with augmentation*

So far, we have implemented the blocks in Figure 2-6 that are green. In this variation, we focus on the block highlighted in blue (Think Augmentation), which we intentionally kept apart. This will help us gauge the impact of the augmentation technique.

Let's look at the augmentation code block:

```
train_transform = transforms.Compose([

    transforms.Resize(224),
    transforms.CenterCrop(224),
    transforms.ColorJitter(brightness=0.10, contrast=0.1,
    saturation=0.10, hue=0.1),
    transforms.RandomHorizontalFlip(),
    transforms.RandomRotation(10),
    transforms.ToTensor(),
    transforms.Normalize([0.485, 0.456, 0.406],
                         [0.229, 0.224, 0.225])
])

test_transform = transforms.Compose([
    transforms.Resize(224),
    transforms.CenterCrop(224),
    transforms.ToTensor(),
    transforms.Normalize([0.485, 0.456, 0.406],
                         [0.229, 0.224, 0.225])
])
```

The augmentation processes are incorporated in the compose functions themselves and thus we don't need to change the data loader portion of the code. We will reuse the exact code process from the compose functions and run it on the iterations.

With data augmentation, we reached an accuracy of 80% on the fifth epoch, but then accuracy drops off during the training process. This shows the erratic nature of the training process. There is a peak in the loss. The plot of the training accuracy shows these fluctuations as well. This is due to the augmentations in the data and the model not being able to figure out the changes efficiently.

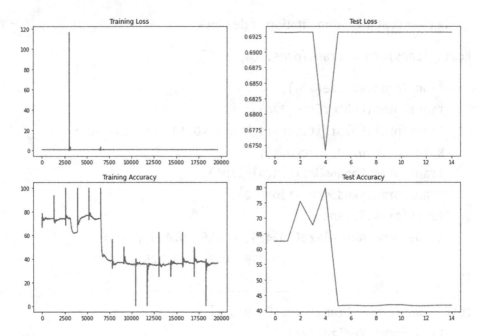

Figure 2-7. *Outputs from the augmentation training pipeline*

Figure 2-7 shows these high accuracies dropping to around 42%. When we compare this workflow accuracy with the earlier version's results, we see that the workflow works much better with data augmentation.

The accuracy is still low on saturation and the model behavior can't be considered stable given the fluctuation. We need to make further changes and look for a better and more stable model.

The Third Variation of the Model

In this section, we consider the established workflow and make amends to it. The model architecture is running with 11 convolution blocks and three max pooling layers. There can be variations in the distributional changes in the layers, also known as internal covariate shifts. We can now try to apply batch normalization in the network architecture.

Batch normalization, as discussed in the earlier chapter, modulates the distributions of the passing input within the layers. The change in distributions has a cascading effect on all the layers preceding it.

We will try to use batch normalization after every layer definition and across all the channels from the block. If we have a convolution block with 16 output channels, that means we need batch normalization on all 16 channels. So far we have accomplished all the functionalities in the computer vision classifier model, as shown in Figure 2-8. We will add batch normalization to the top of this workflow and keep everything carried forward from the last iteration. The green blocks in the figure are already done; we will incorporate the blue block changes (Batch Normalization).

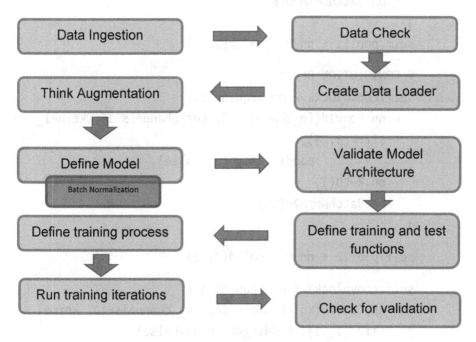

Figure 2-8. *Image classification pipeline with batch normalization*

Let's look at the code block for this model:

```python
class Net(nn.Module):
    def __init__(self):
        super(Net, self).__init__()
        # Input Block
        self.convblock1 = nn.Sequential(
            nn.Conv2d(in_channels=3, out_channels=8, kernel_
            size=(3, 3),
                        padding=0, bias=False),
            nn.ReLU(),
            nn.BatchNorm2d(8)
        )
        self.pool11 = nn.MaxPool2d(2, 2)

        # CONVOLUTION BLOCK 1
        self.convblock2 = nn.Sequential(
            nn.Conv2d(in_channels=8, out_channels=16, kernel_
            size=(3, 3),
                        padding=0, bias=False),
            nn.ReLU(),
            nn.BatchNorm2d(16)
        )

        self.pool22 = nn.MaxPool2d(2, 2)

        self.convblock3 = nn.Sequential(
            nn.Conv2d(in_channels=16, out_channels=10, kernel_
            size=(1, 1), padding=0, bias=False),

            nn.ReLU(),
            nn.BatchNorm2d(10),
        )
        self.pool33 = nn.MaxPool2d(2, 2)
```

```python
# CONVOLUTION BLOCK 2
self.convblock4 = nn.Sequential(
    nn.Conv2d(in_channels=10, out_channels=10, kernel_
    size=(3, 3), padding=0, bias=False),
    nn.ReLU(),
    nn.BatchNorm2d(10)
)

self.convblock5 = nn.Sequential(
    nn.Conv2d(in_channels=10, out_channels=32, kernel_
    size=(1, 1), padding=0, bias=False),

    nn.ReLU(),
    nn.BatchNorm2d(32),

)

self.convblock6 = nn.Sequential(
    nn.Conv2d(in_channels=32, out_channels=10, kernel_
    size=(1, 1), padding=0, bias=False),
    nn.ReLU(),
    nn.BatchNorm2d(10),

)

self.convblock7 = nn.Sequential(
    nn.Conv2d(in_channels=10, out_channels=10, kernel_
    size=(3, 3), padding=0, bias=False),
    nn.ReLU(),
    nn.BatchNorm2d(10)

)
```

```python
        self.convblock8 = nn.Sequential(
            nn.Conv2d(in_channels=10, out_channels=32, kernel_
            size=(1, 1), padding=0, bias=False),

            nn.ReLU(),
            nn.BatchNorm2d(32)
        )

        self.convblock9 = nn.Sequential(
            nn.Conv2d(in_channels=32, out_channels=10, kernel_
            size=(1, 1), padding=0, bias=False),
            nn.ReLU(),
            nn.BatchNorm2d(10)

        )

        self.convblock10 = nn.Sequential(
            nn.Conv2d(in_channels=10, out_channels=14, kernel_
            size=(3, 3), padding=0, bias=False),
            nn.ReLU(),
            nn.BatchNorm2d(14)

        )

        self.convblock11 = nn.Sequential(
            nn.Conv2d(in_channels=14, out_channels=16, kernel_
            size=(3, 3), padding=0, bias=False),
            nn.ReLU(),
            nn.BatchNorm2d(16)

        )

        # OUTPUT BLOCK
        self.gap = nn.Sequential(
            nn.AvgPool2d(kernel_size=4)
        )
```

```python
        self.convblockout = nn.Sequential(
            nn.Conv2d(in_channels=16, out_channels=2, kernel_
            size=(4, 4), padding=0, bias=False),

        )

    def forward(self, x):
        x = self.convblock1(x)
        x = self.pool11(x)
        x = self.convblock2(x)
        x = self.pool22(x)
        x = self.convblock3(x)
        x = self.pool33(x)
        x = self.convblock4(x)
        x = self.convblock5(x)
        x = self.convblock6(x)
        x = self.convblock7(x)
        x = self.convblock8(x)
        x = self.convblock9(x)
        x = self.convblock10(x)
        x = self.convblock11(x)
        x = self.gap(x)
        x = self.convblockout(x)

        x = x.view(-1, 2)
        return F.log_softmax(x, dim=-1)
```

Once we add batch normalization to the model block, the accuracies move up the ladder. The test accuracy reached its max in the tenth epoch, where it was almost 90%. Following that, the accuracies stayed around 85% until the fifteenth epoch. This is a huge improvement in being able to understand the difference between the classes.

There is one more thing to check. For the same processor, we expect
the time consumption per epoch to be higher, but it should not be
significant enough to cause trouble, given the amount of increase in
accuracy it is handing over. Let's look at the model definition as described
by the summary function from torch.

```
----------------------------------------------------------------
        Layer (type)             Output Shape          Param #
================================================================
           Conv2d-1        [-1, 8, 222, 222]              216
            ReLU-2         [-1, 8, 222, 222]                0
       BatchNorm2d-3       [-1, 8, 222, 222]               16
        MaxPool2d-4        [-1, 8, 111, 111]                0
           Conv2d-5       [-1, 16, 109, 109]            1,152
            ReLU-6        [-1, 16, 109, 109]                0
       BatchNorm2d-7      [-1, 16, 109, 109]               32
        MaxPool2d-8         [-1, 16, 54, 54]                0
           Conv2d-9         [-1, 10, 54, 54]              160
           ReLU-10          [-1, 10, 54, 54]                0
      BatchNorm2d-11        [-1, 10, 54, 54]               20
       MaxPool2d-12         [-1, 10, 27, 27]                0
          Conv2d-13         [-1, 10, 25, 25]              900
           ReLU-14          [-1, 10, 25, 25]                0
      BatchNorm2d-15        [-1, 10, 25, 25]               20
          Conv2d-16         [-1, 32, 25, 25]              320
           ReLU-17          [-1, 32, 25, 25]                0
      BatchNorm2d-18        [-1, 32, 25, 25]               64
          Conv2d-19         [-1, 10, 25, 25]              320
           ReLU-20          [-1, 10, 25, 25]                0
      BatchNorm2d-21        [-1, 10, 25, 25]               20
          Conv2d-22         [-1, 10, 23, 23]              900
           ReLU-23          [-1, 10, 23, 23]                0
```

BatchNorm2d-24	[-1, 10, 23, 23]	20
Conv2d-25	[-1, 32, 23, 23]	320
ReLU-26	[-1, 32, 23, 23]	0
BatchNorm2d-27	[-1, 32, 23, 23]	64
Conv2d-28	[-1, 10, 23, 23]	320
ReLU-29	[-1, 10, 23, 23]	0
BatchNorm2d-30	[-1, 10, 23, 23]	20
Conv2d-31	[-1, 14, 21, 21]	1,260
ReLU-32	[-1, 14, 21, 21]	0
BatchNorm2d-33	[-1, 14, 21, 21]	28
Conv2d-34	[-1, 16, 19, 19]	2,016
ReLU-35	[-1, 16, 19, 19]	0
BatchNorm2d-36	[-1, 16, 19, 19]	32
AvgPool2d-37	[-1, 16, 4, 4]	0
Conv2d-38	[-1, 2, 1, 1]	512

==

Total params: 8,732
Trainable params: 8,732
Non-trainable params: 0

Input size (MB): 0.57
Forward/backward pass size (MB): 16.86
Params size (MB): 0.03
Estimated Total Size (MB): 17.46

We can see that the model size has increased from 12.23MB to 17.46MB when we add batch normalization. The number of trainable parameters also went up from 8396 to 8732. From our understanding of the batch normalization concepts, we can see that there will be an increase of two trainable parameters per channel. There should not be any batch normalization or dropouts in the last layer of the model.

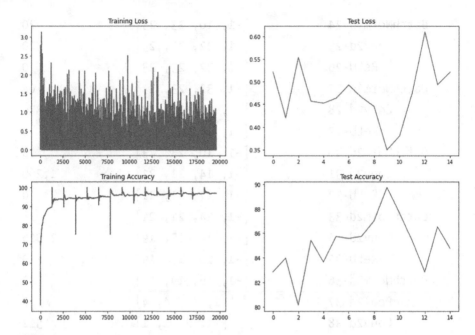

Figure 2-9. *Output from batch normalization pipeline*

Figure 2-9 shows the accuracies and losses from the model training process. We can see from the training loss plot that the fluctuations have reduced significantly from the version where we didn't apply batch normalization. The loss from the test stabilized in the 0.35 to 0.60 region. This loss seems stable on a scale, but we can try one more thing to stabilize this further.

Let's move to one final addition to our methods and check for improvements.

The Fourth Variation of the Model

We will now try to apply regularization and see the difference in losses between the testing and training datasets. We started with a base model and then added augmentations to the training set. After augmentation, we tried to run batch normalization on top of the changes and got good results. For this variation, we use the same workflow as in the third variation and append the regularizer to it.

Figure 2-10 shows the workflow with the completed tasks in green and the pending task (Regularization) in blue.

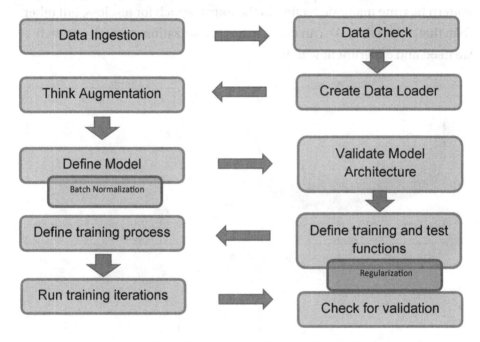

Figure 2-10. *Image classification pipeline with regularization*

The code block for regularizer needs to go within the train function, after we have computed the loss.

```
loss = F.nll_loss(y_pred, target)
        l1 = 0
        for p in model.parameters():
            l1=l1+p.abs().sum()
        loss = loss+lambda1*l1
```

The code block described how we can append the regularizer parameters to the train function and get our training in place. Let's analyze the results from this approach.

Figure 2-10 shows the pattern of train losses and test losses. If we compare this to Figure 2-11, we see that the losses fluctuate less. The pattern has one major deflection in the fourth epoch for test loss, but other than that, it is stable. We can change the regularization strength as much we need and experiment with it.

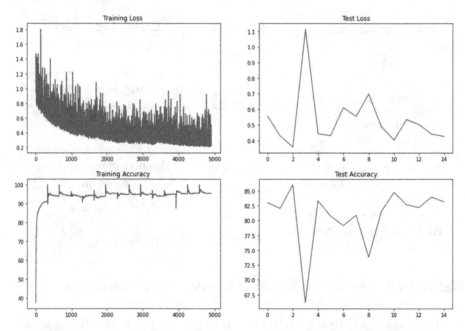

Figure 2-11. *Output from the classification pipeline using regularization*

Summary

We started this chapter by defining a base model and running iterations on the data. We covered some basic augmentation techniques that help create a similar distribution as the test might bring in the future. This helps in building and training a robust model. We explored normalization and regularization to increase the accuracies and stability of the model.

In the next chapter, we look into object detection frameworks, which are based on the concepts learned in this chapter. The image classification networks form the basis of various object detection networks.

CHAPTER 3

Building an Object Detection Model

Object detection is one of the most sought-after skills these days. An image can have multiple classes. In addition, classifying an object solves just part of the problem. The other part lies in the localization of the object. Object detection helps identify the class location of an image with a bounding box. The bounding box can be further processed for various sub-tasks. As an example, think about what a traffic cam needs to detect and identify cars.

The traffic cam needs to detect the car and the license plate and then read the number from the plate to identify the owner. This is not a simple problem to solve. We need annotated registration data. A simple classification convolutional neural network model won't work. We need to get the bounding box of the plate and search for alphanumeric characters, using a series of data cleaning, denoising, and super-resolution steps.

There has been great advancement in the field of object detection lately. Of the numerous object detection approaches, we can break the journey into the pre-2012 era or pre AlexNet era and the post-2012 era. The pre-2012 era includes multiple object detection algorithms such as HOG, Haar cascades, some variations of SIFT, SURF, etc. The post-2012 era includes RCNN, Fast RCNN, Faster RCNN, YOLO, Single Shot Detector, etc.

© Akshay Kulkarni, Adarsha Shivananda, and Nitin Ranjan Sharma 2022
A. Kulkarni et al., *Computer Vision Projects with PyTorch*,
https://doi.org/10.1007/978-1-4842-8273-1_3

We will briefly go through Haar Cascades from the pre-2012 era to establish the context of machine learning-based object detection techniques. To start with Haar Cascades, we need to use features from images rather than the more granular pixels. Haar Cascades was introduced in 2001, and even though it has aged, it is still one of the fastest around.

Object Detection Using Boosted Cascade

Boosted Cascade was primarily built for detecting faces, but it can be used for other object detection tasks. It has three parts to it—the integral images, a boosting algorithm to select the features, and a cascade classifier.

To start, the input images need to be converted to what we call integral images. Integral images can be calculated using simple calculations.

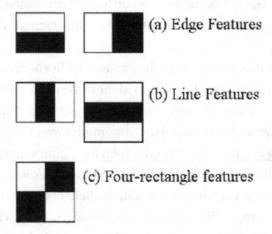

Figure 3-1. *Examples of feature extractors*

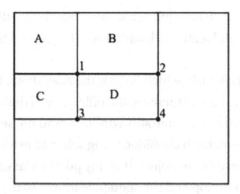

Figure 3-2. Intermediate step

The images in Figure 3-1 show three major types of feature extractors. The first one is the edge extractor followed by line and rectangular feature extractors. Using these extractors, the features need to be selected and the boosting algorithm helps in selecting the necessary ones. The adaptive boosting algorithm presents a set of important features, which aids in faster facial recognition.

Integral image is an intermediate step to get to the feature extraction; it is done by taking the sum of pixels above and to the left of the point at which pixel values are computed, as given in Figure 3-2.

The calculation of the integral images goes like this:

Position 1 = Pixel sum in A rectangle (consider left and above)

Position 2 = Pixel sum A + B

Position 3 = Pixel sum A (above) + C (left)

Position 4 = Pixel sum (4 +1) − sum (2+3)

The extracted features are plotted against the positive samples and negative samples and eventually the best features are chosen. The trained classifier for the positive and negative set of images is formed from the weaker classifiers. For face detection, from a whopping 0.16 million features, the boosting algorithms of a series of weak classifiers help in identifying 6,000 useful features. Eventually, a Cascade classifier helps to detect the class.

The *attentional cascade*, which is what they call them, helps reduce the computational time and increase the efficiency of the detector. The image is segregated into multiple sub-windows, and the sequential weak classifier acts on them. Each classifier, using selected features, tries to check for the presence of the object. If at any point a classifier fails, all subsequent classifiers stop and the sequence moves to the next sub-window, and so on. The detection succeeds if all the classifiers can vote on the presence of the required object and get the bounding box.

Let's go through a sequence of Python code to use the existing model to detect face and eyes.

Import the packages as follows:

```
import cv2
import gc
```

The following function will get the input frame from the camera and scale it for the model. Since color images would not make any difference, a grayscale image is considered. First, the faces are detected, and then for each face, the eyes are located with the help of the other eye detector.

Here's the function to process the cascades of the face and eyes:

```
def detect_face_eye(frame):
    ## normalization and convert color to gray scale
    frame_to_gray = cv2.equalizeHist(cv2.cvtColor(frame, cv2.
    COLOR_BGR2GRAY))
    ## application should be able to different scales of image
    detected_faces = face_cascade.detectMultiScale(frame_to_gray)
    for (x,y,w,h) in detected_faces:
        center_face = (x + w//2, y + h//2)
```

```
## draw an ellipse
frame = cv2.ellipse(frame, center_face, (w//2, h//2),
0, 0, 360, (125, 125, 125), 6)
face_regionofinterest = frame_to_gray[y:y+h,x:x+w]
#detect eyes - for each detected face
## similar multiscale operations
detected_eyes = eyes_cascade.detectMultiScale(face_
regionofinterest)
for (x2,y2,w2,h2) in detected_eyes:
    center_eye = (x + x2 + w2//2, y + y2 + h2//2)
    radius = int(round((w2 + h2)*0.25))
    ## draw a circle
    frame = cv2.circle(frame, center_eye, radius, (255,
    255, 255 ), 4)
cv2.imshow('--Face Detection--', frame)
```

The models can be found in the GitHub repository provided by the
open-cv moderators at https://github.com/opencv/opencv/tree/master/
data/haarcascades. This code uses two models, one for detecting faces and
the other for eyes. However, there are other models present in the repository
that you can experiment with. The function will also access the camera
peripherals attached to the system and use them to scan for faces.

Run the function to enable the face and eye sensing processes:

```
## saved xml paths
face_cascade_name = r' ..\chapter 3\frontal_face_alt.xml'
eyes_cascade_name = r' ..\chapter 3\eye_cascade_model.xml'
## initialize the cascade for detection
face_cascade = cv2.CascadeClassifier()
eyes_cascade = cv2.CascadeClassifier()
## load the cascades face followed by eyes
face_cascade.load(cv2.samples.findFile(face_cascade_name))
eyes_cascade.load(cv2.samples.findFile(eyes_cascade_name))
```

```
camera_device = 0
## enable video processing
capture_cam_img = cv2.VideoCapture(camera_device)
## enable classifier to operate on the face
if capture_cam_img.isOpened :
    while True:
        ret, frame = capture_cam_img.read()
        detect_face_eye(frame)
        ## shut down cv video sensing when ESC is pressed
        if cv2.waitKey(10) == 27:
            cv2.destroyAllWindows()
            gc.collect()
            break
```

Now that we have some idea about structured object detection, we can move to another advanced object detection technique, called R-CNN.

R-CNN

For a long time, objects were separated by image segmentation. Eventually, the hierarchical nature of images posed a bottleneck for developers. Consider if we were trying to locate a person in a car in traffic. An exhaustive search mechanism could be used to scan through each car to find exactly where the person was, but the computation requirement would too high to be useful.

Amid all these problems, the paper entitled "Selective Search for Object Recognition" tried to solve the problem of generating object locations. It uses the best of both worlds—segmentation and exhaustive searching. The following steps show how the selective search mechanism works.

Here's how the algorithm works:

1. The algorithm uses efficient graph-based segmentation to generate the initial regions.

2. The second stage tries to group similar regions to generate segments in the input images. For all the regions created, similarity scores are calculated across all the neighboring elements. The two most similar regions are grouped together and scores are recalculated. This process is repeated until the entire image is covered by the operations.

3. The selection process is convoluted. It uses multiple strategies to get similar regions together. If two regions are being combined, the features of the regions can be propagated through the hierarchy.

4. The selection criteria depends on complementary color spaces, where it indulges in a hierarchical grouping algorithm in a variety of spaces, including finding complementary color spaces. Overall, four fast and efficient strategies help the algorithm.

The main goal of this algorithm is to find diverse yet complementary features to group regions under their strategies.

Precursors in the field of object detection are HOG (Histogram of Oriented Gradients) and SIFT. When the complexity of the visual tasks was acknowledged, a different approach was developed.

The Region Proposal Network

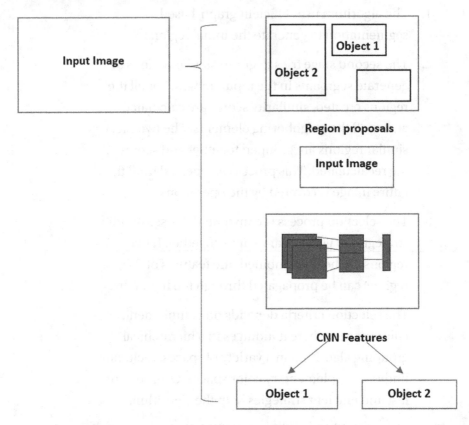

Figure 3-3. *Object detection via region proposal*

Figure 3-3 describes the various steps for object detection in region proposal networks. Let's go through the important steps:

1. Generating segmentation and multiple candidate regions.

2. Using a greedy learning algorithm to combine similar regions into larger ones recursively.

3. Sending the images with the proposals to the convolutional neural network architecture set out for classifying objects.

4. In the case of AlexNET, which is used in standard R-CNN, 227 x 227 is used as the shape of the image.

5. Around 2000 regions are sent to the AlexNET and 4096 vectors are passed on.

6. The extracted features are evaluated against an SVM trained in specific classes.

7. After all the regions are scored, non-max suppression runs on the classified regions. It eliminates regions with IOU to pave the way for regions with greater than the threshold value and a higher region cover.

Interestingly, as the algorithm locates 2000 regions of interest, it generates warped image content from the proposed regions. Since the convolutional blocks need fixed dimensions, the information is warped in space and passed on.

Each of these regions is then classified by support vector machines. Additionally, the algorithm will perform a regression which will rectify or predict any offset to the bounding box predicted in the first place. Before going to the next step, let's review two important concepts that are used multiple times here onward.

- **Non-maximum suppression.** In object detection algorithms, a scenario often arises wherein there are multiple bounding boxes overlapped around one single object. A classifier is often asked to generate probability scores around the regions of interest that are different sizes. To counter the problem of selecting one best bounding box, the algorithm uses classification information and the percentage of coverage on the object.

- **Intersection over Union (IoU).** Used to select the bounding box that has the most similarity with the ground truth. When we are dealing with image classification, we try to map images to their respective classes. Likewise, for object detection, there needs to be a manual intervention of drawing bounding boxes to locate separate objects and classes. The equation gives the ratio of intersection to the union.

The formula for IoU is given by:

$$IoU = \frac{(Bounding\ Box\ 1) \cap (Bounding\ Box\ 2)}{(Bounding\ Box\ 1) \cup (Bounding\ Box\ 2)}$$

Figure 3-4a shows an image with two overlapping bounding boxes, one being the ground truth and the other the predicted truth. Figure 3-4b shows the area aggregated by both of them. These two aspects balance each other out to get the maximum coverage of the ground truth.

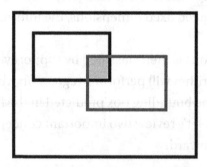

Figure 3-4a. Bounding box interaction

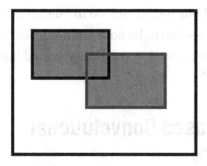

Figure 3-4b. *Bounding box union*

Overall, this algorithm can handle a lot of issues related to object detection and was one of the most revolutionary algorithms when it was published. But it is not without its flaws. Let's dive into some of its glaring flaws now:

- With the help of complex image processing techniques, the model will generate 2000 regions of interest. All of these need to be run by a support vector machine for classification. There is a huge amount of computation involved in this process.

- Most of the algorithms take a lot of time to classify and process images when they are predicting. If we deal with real-time solutions, it is close to impossible to use this model as an algorithm.

- The training is happening in the convolutional section; the classifier and the regression are correcting the bounding box parameters.

- In the initial section of the algorithm, a selective search mechanism is used to segment similar regions and collectively generate regions of interest. The entire process is based on the collocation of complex image processing techniques. There is no learning involved in the process, thus the scope for improvement is minimal.

Amid all the problems R-CNN solved in object detection, it left behind a series of issues that needed to be addressed. An advancement to region-based object detection network came along, named fast region-based convolutional network.

Fast Region-Based Convolutional Neural Network

To set the context, if there is an image on which object detection should work, as per the simple region-based convolutional neural network, it will generate the regions of interest on top of simple images. The combinations will be huge. But if we can reduce the image to a smaller size in terms of (x,y), we can still get the portion of the image with the correct object information. Eventually, this boils down to how we are conveying the information to the subsequent layers and the cost function. Fast R-CNN leads to faster operations.

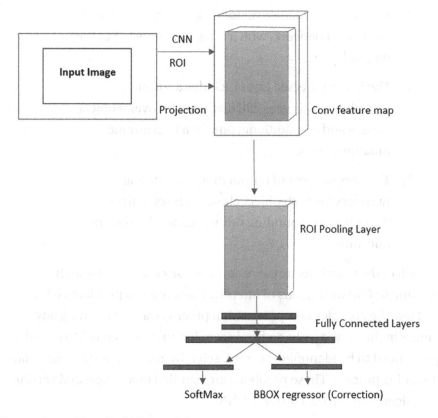

CNN

Input Image

ROI

Projection   Conv feature map

ROI Pooling Layer

Fully Connected Layers

SoftMax   BBOX regressor (Correction)

Figure 3-5. *Fast R-CNN architecture*

Figure 3-5 depicts the workflow based on one region of interest. The architecture suggests that a convolution operation is performed on the input data, thus reducing the computations.

The processes involved in the fast region-based object detection are as follows:

1. A feature map is created from multiple convolutions and pooling operations.

2. Since the fully connected network will want a vector of fixed dimensions, the region of interest's pooling layer extracts a fixed-length vector.

3. Each of these feature vectors is input into the fully
    connected network, which is again connected to the
    output layers.

4. The first connected layer includes a softmax
    probability estimates calculation, with over n object
    classes and an additional one for a background or
    unknown class.

5. The second layer of output predicts four real
    numbers for each object class. Each set defines
    the corrected bounding box values for the class in
    question.

Each of these architectures we covered uses a selective search
algorithm to find the regions of interest. There are two problems with
this. First, the complex computer vision process was not learning any
changes in the data, as it has a fixed set of instructions related to how the
regions need to be identified. Second, selective search is a slow and time-
consuming process. These problems are handled in an upgraded version
of the algorithm, called a faster R-CNN.

How the Region Proposal Network Works

An extended idea came along, which was suggested and implemented as
well, which used a neural network to predict the region proposals without
a selective search mechanism. The region proposal network came along to
help identify the bounding boxes in the images and then the same block is
sent out to the convolutional neural network for feature maps.

Eventually, the loss functions are trained on the feature maps and the network weights are adjusted to accommodate the training. Let's go through this process step by step:

1. In the first step, the input image is passed onto a convolutional block to generate the convolutional feature maps.

2. A sliding window is used on the feature map for each location, by the region proposal network.

3. For each location, nine anchor boxes are used with three different scales and three aspect ratios (1:1, 1:2, 2:1), which helps generate the region proposals.

4. The classification layer tells the output whether there is an object present in the anchor boxes.

5. The regression layer indicates the coordinates for the anchor boxes.

6. The anchor boxes are passed to the region of interest's pooling layer of the Fast R-CNN architecture.

We have used a neural network to learn where the region proposals are and how they can be tuned based on the data. This also makes the process much faster than the one we learned about earlier. Figure 3-6 shows the architecture from the primary research paper of Faster R-CNN.

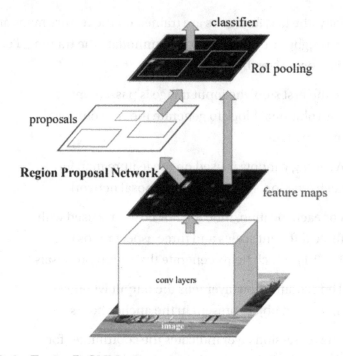

Figure 3-6. *Faster R-CNN summary*

The network has a novel idea of a region proposal network that learns the bounding boxes and can generalize them. It has three main types of networks:

- **Head:** Can be a ResNet architecture, which serves the purpose of generating feature maps.

- **Region Proposal Network:** Generates the region of interest for the classifier and the regressor.

- **Classification Network/ Regression Network:** Handles the classification of objects and objectness or the correctness of the bounding box coordinates.

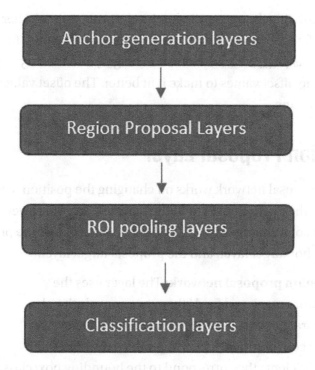

Figure 3-7. *Faster R-CNN flowchart*

Figure 3-7 depicts the basic layers of Faster R-CNN. Let's go into the details, which will enhance the development of the layers.

The Anchor Generation Layer

This layer produces a series of bounding boxes with different sizes and aspect ratios to cover most of the image regions. The bounding boxes or the anchor boxes will encompass images and their objects. These boxes, however, will be content-agnostic and the same throughout, and eventually the region proposal network will work on them and identify which of them is a better bounding box. Little tweaks will lead to a better bounding box.

Since predicting these coordinates has its problems, another approach is to take a reference box as a standard for the bounding box. Take a reference box as (X_{center}, Y_{center}, width, and height) and then try to predict and correct the offset values to make it fit better. The offset values are for all four parameters.

The Region Proposal Layer

The region proposal network works on changing the position, width, and height of the anchor boxes to fit the object better. This layer can be considered a combination of the region proposal network, the proposal layer, the anchor target layer, and the proposal target layer.

- **Region proposal network:** The layer uses the feature maps and feeds them to a convolutional neural network. The output is then passed to two 1x1 convolutional layers to produce the regression coefficients the correspond to the bounding box, class scores, and probabilities.

- **Proposal layer:** This layer takes the numerous anchor boxes and reduces them to an appropriate number by taking help from non-max suppression based on the foreground scores. It also changes the coordinates of the bounding boxes using the coefficients generated by the region proposal network.

- **Anchor target layer:** This helps in the selection of the anchor boxes that help the RPN differentiate between the foreground and the background.

The loss function for RPN is a combination of classification and regression loss.

$$loss = Regression\ Loss_{Bounding\ Box} + Classification\ Loss$$

Overall, Faster R-CNN has a convolutional neural network-based image feature extractor and region proposal network to generate the regions of interest. We have ROI pooling to get the images to fixed dimensions for the next layers and finally to the classification and regression layers. This helps the anchor boxes be a better fit and be good enough to differentiate between the foreground and background.

Mask R-CNN

On top of what Faster R-CNN has already achieved, this is an extension in terms of predicting masks on the detected objects. There are two more convolutional neural networks after the ROI pooling layer to add masks. This also establishes ROI Align, which helps better align the extracted features with the inputs and avoid the warping that used to happen in Faster R-CNN. It uses bilinear interpolation to get the exact or near-perfect values of the input regions.

An important step that came out of all these object detection methods was the use of anchor boxes. YOLO extends this idea with some additional changes, which we will be looking into.

Prerequisites

- *Annotation*. In a classification problem, images need to sorted or arranged following the classes they belong to. Similarly, in an object detection problem, the images need to be marked with appropriate bounding boxes, often called the ground truth. The bounding box will serve the purpose of suggesting the coordinates of the object and the classes being enclosed. Figure 3-8 shows one instance of images that have been annotated to train a classifier where the birds are located.

Generally, annotations are done manually, sometimes multiple times, to get the ground truth unbiased. However, if we are doing it for practice like now, we can use any open source data for experimentation purposes.

***Figure 3-8.** Birds with annotations*

- **GPU preferred.** When working in computer vision tasks, which requires training and running inferences, it is advised to get CUDA-enabled GPU cores for faster processing.

- **Torch framework installed with CUDA capabilities.** We also need PyTorch installed on the system, as discussed in the previous chapter.

YOLO

There were huge requirements for object detection algorithms that will help inferences in real-time. Faster R-CNN came very close to that, coping with the 2000 bounding box predictions and the traditional computer vision method. It had significant improvements over its predecessors but still had scope for improvement.

In came the revolutionary algorithm YOLO, which made the detection of objects in 45 frames per second (TITAN X). The earlier models spent too much time training and predicting at different levels, such as the anchor generation layer, the region proposal layer, classification, and bounding box correction. YOLO, on the other hand, tries to get one convolutional neural network block to predict bounding boxes and classes, thus reducing computation time. It has a more generalized way of training and considering information from the entire image rather than piecing it together. Finally, it supersedes other predecessors tat are trying to do the same thing.

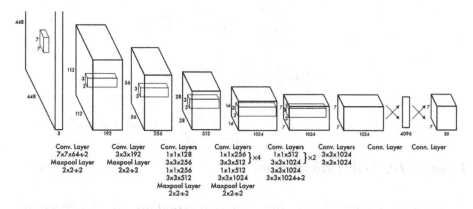

Figure 3-9. *YOLO architecture*

Figure 3-9 shows the architecture in YOLO, which is inspired by the GoogleNet architecture for image classification. The input layer shows a dimension of 448x448x3. The network has 24 convolutional layers and max pooling layers in batches with two fully connected layers.

The training process is fairly expensive, so training an object detection model from scratch needs to be under good governance. The training for this given architecture is done in two ways. First, the model is trained on ImageNet data, with the first 20 convolutional layers, an average pooling to match the dimensions with a fully connected network. This block is trained for a week to get an 88% accuracy.

This pretrained network is added with four convolutional layers and two fully connected layers to get the final detected object. The input dimensions are also increased from 224x224 to 448x448, which helps in the detection abilities. The final layer predicts the classification scores and the bounding box coordinates. The width and height of the bounding boxes are normalized.

$$\lambda_{\text{coord}} \sum_{i=0}^{S^2} \sum_{j=0}^{B} \mathbb{1}_{ij}^{\text{obj}} \left[(x_i - \hat{x}_i)^2 + (y_i - \hat{y}_i)^2 \right]$$

$$+ \lambda_{\text{coord}} \sum_{i=0}^{S^2} \sum_{j=0}^{B} \mathbb{1}_{ij}^{\text{obj}} \left[\left(\sqrt{w_i} - \sqrt{\hat{w}_i} \right)^2 + \left(\sqrt{h_i} - \sqrt{\hat{h}_i} \right)^2 \right]$$

$$+ \sum_{i=0}^{S^2} \sum_{j=0}^{B} \mathbb{1}_{ij}^{\text{obj}} \left(C_i - \hat{C}_i \right)^2$$

$$+ \lambda_{\text{noobj}} \sum_{i=0}^{S^2} \sum_{j=0}^{B} \mathbb{1}_{ij}^{\text{noobj}} \left(C_i - \hat{C}_i \right)^2$$

$$+ \sum_{i=0}^{S^2} \mathbb{1}_{i}^{\text{obj}} \sum_{c \in \text{classes}} (p_i(c) - \hat{p}_i(c))^2$$

Figure 3-10. *YOLO loss function*

Figure 3-10 shows the loss function used to optimize the classification and regression. For each of the five anchor boxes, there is an objectness score, four coordinates corresponding to the normalized bounding box, and the top class probabilities or scores. These changes have worked quite well but needed further refinement. Let's look at the second version update and version 3, which was one of the most popular models.

YOLO V2/V3

The changes in YOLO were remarkable and the second version fine-tunes the approach to a more efficient level. Here are a few key pointers that were addressed in the second version:

- The convolutional layers ran deep, so there was always a chance of a vanishing gradient or exploding gradient. Batch normalization was added to help with the internal covariate shift in the learning.

- It predicts class and objectness for every anchor box.

- The network also predicts five bounding boxes and five coordinates for each bounding box.

- A major architectural change happened when it removed the fully connected layers and replace them with anchor boxes to predict bounding boxes.

- These anchor boxes are determined with the help of clustering on the bounding boxes of ground truth.

Even after multiple changes, the researchers found that a few more changes could improve the accuracies. They made the required changes and named this version YOLO V3. It was arguably one of the most sought-after object detection architectures. Whereas YOLO was using the softmax layer to get the final classification scores, YOLO V3 resorted to using individual logistic regression or multi-label classification on the inputs. Interestingly, it also removes the pooling layer and instead uses 3x3 with a stride of 2 to reduce dimensionality.

The architecture also made changes to the loss function, with three major predictions coming out—coordinates of the bounding boxes, the objectness value, and the class scores. The most popular backbone in the YOLO V3 architecture is Darnet-53, which is a 53-layered architecture of convolutional blocks, as shown in Figure 3-11. It uses the residual implementation with 3x3 and 1x1 convolutional layers to get the features for detection and classification. Overall, the changes made a huge impact on the accuracy and optimization of the architecture.

	Type	Filters	Size	Output
	Convolutional	32	3 × 3	256 × 256
	Convolutional	64	3 × 3 / 2	128 × 128
	Convolutional	32	1 × 1	
1×	Convolutional	64	3 × 3	
	Residual			128 × 128
	Convolutional	128	3 × 3 / 2	64 × 64
	Convolutional	64	1 × 1	
2×	Convolutional	128	3 × 3	
	Residual			64 × 64
	Convolutional	256	3 × 3 / 2	32 × 32
	Convolutional	128	1 × 1	
8×	Convolutional	256	3 × 3	
	Residual			32 × 32
	Convolutional	512	3 × 3 / 2	16 × 16
	Convolutional	256	1 × 1	
8×	Convolutional	512	3 × 3	
	Residual			16 × 16
	Convolutional	1024	3 × 3 / 2	8 × 8
	Convolutional	512	1 × 1	
4×	Convolutional	1024	3 × 3	
	Residual			8 × 8
	Avgpool		Global	
	Connected		1000	
	Softmax			

Figure 3-11. *Darknet 53 architecture*

Let's look at some code that uses a saved model and tunes it for a custom dataset. Why are we not training it from scratch? These are all heavyweight models, and we don't always have enough GPU capacity to train from scratch. Secondly, using the trained weights and modifying them accordingly is a learning experience. A term we will keep repeating is *transfer learning.*

Project Code Snippets

The code snippet is adapted from the original creators of YOLO, and all source code credits go to Joseph Redmon and Ali Farhadi. Even though training from scratch is fairly complex, we can try to use existing open source models to do a transfer learning on this data. We can also use the existing models to run inference on our data if the classes on which the original model were trained closely resemble the ones we are using.

The folder setup needs to follow the original creators' since we will be using the saved models to custom train our data. As shown in Figure 3-12, for any variations, the path should be corrected according to the configuration files—under data.

Figure 3-12. Folder structure for YOLO

Step 1: Getting Annotated Data

Image annotations are one of the most vital prerequisites of an object detection algorithm when we want to train custom data. They help the model with the classification and regression loss functions. They have the ground truth, which is manually resolved. There are multiple open-sourced locations from where we can annotate images. The tool will generally have a marker that will help draw a bounding box of some shape on top of the image. The program will allow the annotations to be downloaded in JSON, CSV, or VOC/COCO format, depending on the model being used. The training and custom data should align.

It's immensely important that annotation is correct and true to the person annotating. Since it is a manual and repetitive task it needs to be as good as it can be. Eventually, the generated file should be downloaded and placed in the data folder. For instance, each image might look like this:

```
0 0.41833333333333333 0.2112676056338028
0.2011111111111111 0.2007042253521127

2 0.43777777777777777 0.3970070422535211
0.11555555555555555 0.15669014084507044

1 0.38722222222222225 0.6813380281690141 0.47
0.4119718309859155
```

Once we have aggregated the new files, we can see how to change the data files. In Figure 3-12, the folders under the data are primarily labels and images. The images have the original images with the same name as the one annotated. The text files need to have the annotation information and be placed in the labels. This will either be a text file or JSON.

Once this is done, we will check the file's custom data file, which needs to be updated with information such as where the train and test file information is stored. We need to project two kinds of information here— the path to the labels and images and the actual images. The custom data file will look something like this:

```
classes=4

train=data/train.txt

valid=data/test.txt

names=data/custom.names
```

This gives relative information about the data and its whereabouts. Once this is done, we need to provide the class names in the custom names file. It will look something like this:

```
hardhat

vest

mask

boots
```

This file lists the class names to the number linked earlier. As mentioned earlier, we need `train.txt` and `test.txt` files with the path to the images. The files should have a relative path to run the training function.

There are other files such as train and test shapes (`train.shapes` and `test.shapes`) that have the shapes for all the files, which we can change according to the input data.

Once all these are done, we have to download the saved weights from the source and the original researchers at `https://pjreddie.com/darknet/yolo/`. There are various options according to the GPU strength of the person working on the project. The weights and the configuration files are tied to each other. So be careful to download the corresponding config file for the weights. With these primary steps, the initial setup is done. Now we move to the next process.

Step 2: Fixing the Configuration File and Training

Another important task is changing the config files as per the requirements and resources. Figure 3-13 shows the first changes in the training and testing configuration. It has provisions for altering the batch size, width, height, channels, momentum, and decay parameters.

Figure 3-13. *Changes in the config file for training/testing*

Important parameters such as learning rates and burn-in are also provided. Apart from these changes, there are changes concerning the classes and the final layer. Since we will be doing a custom train against the original training method, which included 80 classes, this situation can vary. Figure 3-14 shows a few required changes. If the training can happen in the default coco dataset, the original config files can be used.

Figure 3-14. *Changes required in the config for the training/inference pipeline*

All the instances of classes and filters need to be changed in the configuration file. We need to change [filters=255] to filters=(number of classes + 5)x3 in the instances that come before the YOLO layer, as shown in line 640.

After we make these changes, we can move to the training section. Only one job needs to be run.

```
!python train.py --data $PATH/custom.data --batch $num_batches
--cache --epochs $num_epochs -nosave

$num_batches = Number of batches
$num_epochs = Number of epochs of training (Remember this is
transfer learning and we are already using the saved weights)
$path = path to custom data.
```

If we are running out of memory, we either try reducing the model parameters by taking smaller saved models and using them for training, or we reduce the batch size or image resolutions. Whichever method seems easy and appropriate is fine.

The project has too many dependencies and it is advised to take a direct reference and use it to save time and use an optimized version of the code. The training code, model code, and configuration code are inter-connected. the configuration file has a direct impact on the training process and the model setup. Let's look at the Python code for the model definition from the source code used by the researchers.

The Model File

There are standard imports with torchvision and torch functions, which are used multiple times in the code. The parse package is used to get the command-line arguments. The first function that is present in the model file is the create_modules function. Let's go through some important steps, just in case a "train from the scratch" scenario is created.

```
def create_modules(module_defs, img_size):
    # Constructs module list of layer blocks from module
    configuration in module_defs

    img_size = [img_size] * 2 if isinstance(img_size, int) else
    img_size  # expand if necessary
    _ = module_defs.pop(0)  # cfg training hyperparams (unused)
    output_filters = [3]  # input channels
    module_list = nn.ModuleList()
    routs = []  # list of layers which rout to deeper layers
    yolo_index = -1

    for i, mdef in enumerate(module_defs):
        modules = nn.Sequential()

        if mdef['type'] == 'convolutional':
            bn = mdef['batch_normalize']
            filters = mdef['filters']
            k = mdef['size']  # kernel size
            stride = mdef['stride'] if 'stride' in mdef else
            (mdef['stride_y'], mdef['stride_x'])
            if isinstance(k, int):  # single-size conv
                modules.add_module('Conv2d', nn.Conv2d(in_
                channels=output_filters[-1],
                                                       out_channels=
                                                       filters,
                                                       kernel_size=k,
                                                       stride=stride,
                                                       padding=k
                                                       // 2 if
                                                       mdef['pad']
                                                       else 0,
```

```
                                            groups=
                                            mdef['groups']
                                            if 'groups' in
                                            mdef else 1,
                                            bias=not bn))
        else:  # multiple-size conv
            modules.add_module('MixConv2d',
            MixConv2d(in_ch=output_filters[-1],
                                            out_ch=filters,
                                            k=k,
                                            stride=stride,
                                            bias=not bn))

        if bn:
            modules.add_module('BatchNorm2d',
            nn.BatchNorm2d(filters, momentum=0.03, eps=1E-4))
        else:
            routs.append(i)  # detection output (goes into
            yolo layer)

        if mdef['activation'] == 'leaky':  # activation
        study https://github.com/ultralytics/yolov3/
        issues/441
            modules.add_module('activation',
            nn.LeakyReLU(0.1, inplace=True))
            # modules.add_module('activation',
            nn.PReLU(num_parameters=1, init=0.10))
        elif mdef['activation'] == 'swish':
            modules.add_module('activation', Swish())

    elif mdef['type'] == 'BatchNorm2d':
        filters = output_filters[-1]
```

```
        modules = nn.BatchNorm2d(filters, momentum=0.03,
        eps=1E-4)
        if i == 0 and filters == 3:  # normalize RGB image
            # imagenet mean and var https://pytorch.
            org/docs/stable/torchvision/models.
            html#classification
            modules.running_mean = torch.tensor([0.485,
            0.456, 0.406])
            modules.running_var = torch.tensor([0.0524,
            0.0502, 0.0506])

    elif mdef['type'] == 'maxpool':
        k = mdef['size']  # kernel size
        stride = mdef['stride']
        maxpool = nn.MaxPool2d(kernel_size=k,
        stride=stride, padding=(k - 1) // 2)
        if k == 2 and stride == 1:  # yolov3-tiny
            modules.add_module('ZeroPad2d',
            nn.ZeroPad2d((0, 1, 0, 1)))
            modules.add_module('MaxPool2d', maxpool)
        else:
            modules = maxpool

    elif mdef['type'] == 'upsample':
        if ONNX_EXPORT:  # explicitly state size, avoid
        scale_factor
            g = (yolo_index + 1) * 2 / 32  # gain
            modules = nn.Upsample(size=tuple(int(x * g) for
            x in img_size))  # img_size = (320, 192)
        else:
            modules = nn.Upsample(scale_factor=
            mdef['stride'])
```

```
elif mdef['type'] == 'route':  # nn.Sequential()
placeholder for 'route' layer
    layers = mdef['layers']
    filters = sum([output_filters[l + 1 if l > 0
    else l] for l in layers])
    routs.extend([i + l if l < 0 else l for l in layers])
    modules = FeatureConcat(layers=layers)

elif mdef['type'] == 'shortcut':  # nn.Sequential()
placeholder for 'shortcut' layer
    layers = mdef['from']
    filters = output_filters[-1]
    routs.extend([i + l if l < 0 else l for l in layers])
    modules = WeightedFeatureFusion(layers=layers,
    weight='weights_type' in mdef)

elif mdef['type'] == 'reorg3d':  # yolov3-spp-pan-scale
    pass

elif mdef['type'] == 'yolo':
    yolo_index += 1
    stride = [32, 16, 8, 4, 2][yolo_index]  # P3-
    P7 stride
    layers = mdef['from'] if 'from' in mdef else []
    modules = YOLOLayer(anchors=mdef['anchors']
    [mdef['mask']],  # anchor list
                        nc=mdef['classes'],  # number
                        of classes
                        img_size=img_size,  # (416, 416)
                        yolo_index=yolo_index,  #
                        0, 1, 2...
                        layers=layers,  # output layers
                        stride=stride)
```

117

```
            # Initialize preceding Conv2d() bias
            (https://arxiv.org/pdf/1708.02002.pdf section 3.3)
            try:
                j = layers[yolo_index] if 'from' in mdef else -1
                bias_ = module_list[j][0].bias  # shape(255,)
                bias = bias_[:modules.no * modules.na].
                view(modules.na, -1)  # shape(3,85)
                bias[:, 4] += -4.5  # obj
                bias[:, 5:] += math.log(0.6 / (modules.nc -
                0.99))  # cls (sigmoid(p) = 1/nc)
                module_list[j][0].bias = torch.
                nn.Parameter(bias_, requires_grad=bias_.
                requires_grad)
            except:
                print('WARNING: smart bias initialization
                failure.')

        else:
            print('Warning: Unrecognized Layer Type: ' +
            mdef['type'])

        # Register module list and number of output filters
        module_list.append(modules)
        output_filters.append(filters)

    routs_binary = [False] * (i + 1)
    for i in routs:
        routs_binary[i] = True
    return module_list, routs_binary
```

The important steps found in this code are:

1. A sequential model is initialized, which sets the context of the model blocks.

2. The model takes arguments from the command line and gets the variables related to batch normalization, filters, activation function, and convolution.

3. There is an option to save the model as an ONNX version.

After the initial model definitions, we have the YOLO layer class, which uses functions to define the model as per the configuration received. Let's look at the code as provided by the source research.

```
class YOLOLayer(nn.Module):
    def __init__(self, anchors, nc, img_size, yolo_index,
    layers, stride):
        super(YOLOLayer, self).__init__()
        self.anchors = torch.Tensor(anchors)
        self.index = yolo_index  # index of this layer in layers
        self.layers = layers  # model output layer indices
        self.stride = stride  # layer stride
        self.nl = len(layers)  # number of output layers (3)
        self.na = len(anchors)  # number of anchors (3)
        self.nc = nc  # number of classes (80)
        self.no = nc + 5  # number of outputs (85)
        self.nx, self.ny, self.ng = 0, 0, 0  # initialize
        number of x, y gridpoints
        self.anchor_vec = self.anchors / self.stride
        self.anchor_wh = self.anchor_vec.view(1, self.na, 1, 1, 2)
```

```
    if ONNX_EXPORT:
        self.training = False
        self.create_grids((img_size[1] // stride, img_
        size[0] // stride))  # number x, y grid points

def create_grids(self, ng=(13, 13), device='cpu'):
    self.nx, self.ny = ng  # x and y grid size
    self.ng = torch.tensor(ng)

    # build xy offsets
    if not self.training:
        yv, xv = torch.meshgrid([torch.arange(self.ny,
        device=device), torch.arange(self.nx,
        device=device)])
        self.grid = torch.stack((xv, yv), 2).view((1, 1,
        self.ny, self.nx, 2)).float()

    if self.anchor_vec.device != device:
        self.anchor_vec = self.anchor_vec.to(device)
        self.anchor_wh = self.anchor_wh.to(device)

def forward(self, p, out):
    ASFF = False  # https://arxiv.org/abs/1911.09516
    if ASFF:
        i, n = self.index, self.nl  # index in layers,
        number of layers
        p = out[self.layers[i]]
        bs, _, ny, nx = p.shape  # bs, 255, 13, 13
        if (self.nx, self.ny) != (nx, ny):
            self.create_grids((nx, ny), p.device)

        # outputs and weights
        # w = F.softmax(p[:, -n:], 1)  # normalized weights
```

```
w = torch.sigmoid(p[:, -n:]) * (2 / n)
    # sigmoid weights (faster)
# w = w / w.sum(1).unsqueeze(1)
    # normalize across layer dimension

# weighted ASFF sum
p = out[self.layers[i]][:, :-n] * w[:, i:i + 1]
for j in range(n):
    if j != i:
        p += w[:, j:j + 1] * \
            F.interpolate(out[self.layers[j]]
            [:, :-n], size=[ny, nx],
            mode='bilinear', align_corners=False)

elif ONNX_EXPORT:
    bs = 1  # batch size
else:
    bs, _, ny, nx = p.shape  # bs, 255, 13, 13
    if (self.nx, self.ny) != (nx, ny):
        self.create_grids((nx, ny), p.device)

# p.view(bs, 255, 13, 13) -- > (bs, 3, 13, 13, 85)
  # (bs, anchors, grid, grid, classes + xywh)
p = p.view(bs, self.na, self.no, self.ny, self.nx).
permute(0, 1, 3, 4, 2).contiguous()  # prediction

if self.training:
    return p

elif ONNX_EXPORT:
    # Avoid broadcasting for ANE operations
    m = self.na * self.nx * self.ny
    ng = 1 / self.ng.repeat((m, 1))
```

```
        grid = self.grid.repeat((1, self.na, 1, 1, 1)).
        view(m, 2)
        anchor_wh = self.anchor_wh.repeat((1, 1, self.nx,
        self.ny, 1)).view(m, 2) * ng

        p = p.view(m, self.no)
        xy = torch.sigmoid(p[:, 0:2]) + grid  # x, y
        wh = torch.exp(p[:, 2:4]) * anchor_wh
          # width, height
        p_cls = torch.sigmoid(p[:, 4:5]) if self.nc ==
        1 else \
            torch.sigmoid(p[:, 5:self.no]) * torch.
            sigmoid(p[:, 4:5])  # conf
        return p_cls, xy * ng, wh

    else:  # inference
        io = p.clone()  # inference output
        io[..., :2] = torch.sigmoid(io[..., :2]) + self.
        grid  # xy
        io[..., 2:4] = torch.exp(io[..., 2:4]) * self.
        anchor_wh  # wh yolo method
        io[..., :4] *= self.stride
        torch.sigmoid_(io[..., 4:])
        return io.view(bs, -1, self.no), p  # view [1, 3,
        13, 13, 85] as [1, 507, 85]
```

This code defines the YOLO layer, uses the initialization, and sets up everything perfectly for training. The important section of the code is as follows:

1. The YOLO layer is being configured with useful information, such as the number of anchors, the classes, the number of outputs, and the number of classes.

2. The code also sets the grid on the images, which
    is required for anchor boxes. It also sets the
    parameters for the forward propagation.

3. It also lets a provision set the ONNX model.

Finally, the detection model code is placed, and it uses the Darknet
framework to create a highly optimized workflow for object detection.

```python
class Darknet(nn.Module):
    # YOLOv3 object detection model

    def __init__(self, cfg, img_size=(416, 416), verbose=False):
        super(Darknet, self).__init__()

        self.module_defs = parse_model_cfg(cfg)
        self.module_list, self.routs = create_modules(self.
        module_defs, img_size)
        self.yolo_layers = get_yolo_layers(self)
        # torch_utils.initialize_weights(self)

        # Darknet Header https://github.com/AlexeyAB/darknet/
        issues/2914#issuecomment-496675346
        self.version = np.array([0, 2, 5], dtype=np.int32)
          # (int32) version info: major, minor, revision
        self.seen = np.array([0], dtype=np.int64)
          # (int64) number of images seen during training
        self.info(verbose) if not ONNX_EXPORT else None
          # print model description

    def forward(self, x, augment=False, verbose=False):

        if not augment:
            return self.forward_once(x)
        else:  # Augment images (inference and test only)
        https://github.com/ultralytics/yolov3/issues/931
```

```
img_size = x.shape[-2:]  # height, width
s = [0.83, 0.67]  # scales
y = []
for i, xi in enumerate((x,
                        torch_utils.scale_img
                        (x.flip(3), s[0],
                        same_shape=False),
                          # flip-lr and scale
                        torch_utils.scale_
                        img(x, s[1], same_
                        shape=False),  # scale
                        )):
    # cv2.imwrite('img%g.jpg' % i, 255 * xi[0].
    numpy().transpose((1, 2, 0))[:, :, ::-1])
    y.append(self.forward_once(xi)[0])

y[1][..., :4] /= s[0]  # scale
y[1][..., 0] = img_size[1] - y[1][..., 0]  # flip lr
y[2][..., :4] /= s[1]  # scale

# for i, yi in enumerate(y):  # coco small, medium,
large = < 32**2 < 96**2 <
#     area = yi[..., 2:4].prod(2)[:, :, None]
#     if i == 1:
#         yi *= (area < 96. ** 2).float()
#     elif i == 2:
#         yi *= (area > 32. ** 2).float()
#     y[i] = yi

y = torch.cat(y, 1)
return y, None
```

```python
def forward_once(self, x, augment=False, verbose=False):
    img_size = x.shape[-2:]  # height, width
    yolo_out, out = [], []
    if verbose:
        print('0', x.shape)
        str = ''

    # Augment images (inference and test only)
    if augment:  # https://github.com/ultralytics/yolov3/
    issues/931
        nb = x.shape[0]  # batch size
        s = [0.83, 0.67]  # scales
        x = torch.cat((x,
                    torch_utils.scale_img(x.flip(3),
                    s[0]),  # flip-lr and scale
                    torch_utils.scale_img(x,
                    s[1]),  # scale
                    ), 0)

    for i, module in enumerate(self.module_list):
        name = module.__class__.__name__
        if name in ['WeightedFeatureFusion',
        'FeatureConcat']:  # sum, concat
            if verbose:
                l = [i - 1] + module.layers  # layers
                sh = [list(x.shape)] + [list(out[i].shape)
                for i in module.layers]  # shapes
                str = ' >> ' + ' + '.join(['layer %g %s' %
                x for x in zip(l, sh)])
            x = module(x, out)  # WeightedFeatureFusion(),
            FeatureConcat()
        elif name == 'YOLOLayer':
            yolo_out.append(module(x, out))
```

```python
        else:  # run module directly, i.e. mtype
        = 'convolutional', 'upsample', 'maxpool',
        'batchnorm2d' etc.
            x = module(x)

        out.append(x if self.routs[i] else [])
        if verbose:
            print('%g/%g %s -' % (i, len(self.module_list),
            name), list(x.shape), str)
            str = ''

    if self.training:  # train
        return yolo_out
    elif ONNX_EXPORT:  # export
        x = [torch.cat(x, 0) for x in zip(*yolo_out)]
        return x[0], torch.cat(x[1:3], 1)  # scores, boxes:
        3780x80, 3780x4
    else:  # inference or test
        x, p = zip(*yolo_out)  # inference output,
        training output
        x = torch.cat(x, 1)  # cat yolo outputs
        if augment:  # de-augment results
            x = torch.split(x, nb, dim=0)
            x[1][..., :4] /= s[0]  # scale
            x[1][..., 0] = img_size[1]`- x[1][..., 0]
            # flip lr
            x[2][..., :4] /= s[1]  # scale
            x = torch.cat(x, 1)
        return x, p

def fuse(self):
    # Fuse Conv2d + BatchNorm2d layers throughout model
    print('Fusing layers...')
```

```
fused_list = nn.ModuleList()
for a in list(self.children())[0]:
    if isinstance(a, nn.Sequential):
        for i, b in enumerate(a):
            if isinstance(b, nn.modules.batchnorm.
            BatchNorm2d):
                # fuse this bn layer with the previous
                conv2d layer
                conv = a[i - 1]
                fused = torch_utils.fuse_conv_and_
                bn(conv, b)
                a = nn.Sequential(fused, *list(a.
                children())[i + 1:])
                break
    fused_list.append(a)
self.module_list = fused_list
self.info() if not ONNX_EXPORT else None
    # yolov3-spp reduced from 225 to 152 layers

def info(self, verbose=False):
    torch_utils.model_info(self, verbose)
```

These steps use the Darknet framework, which is available at
https://pjreddie.com/darknet/. It is fast and highly optimized for
use in computer vision problems. Apart from this, the model file has
configuration details that seek existing weights for use and other details.
The training file has most of the configurable details, including settings to
deal with the data path, the configuration file path, and other architectural
details. It also sets and freezes the weights on which training is already
done, and it trains only those layers that require training and updating.
With this, we conclude the training process of YOLO.

Summary

Object detection is a difficult process that requires solving multiple tasks at the same time. It needs to be optimized for real-time usage. In this chapter, we explored mechanisms that allow the model to learn the classification and the localization of objects.

It all comes down to the fact that a machine, if allowed, can wield extensive power to learn constraints. Object detection algorithms can be used in day-to-day work, including with self-driving cars, traffic cameras, security drones, and many more use cases.

In the next chapter, we look at image segmentation, which is similar to the process we discussed earlier. Image segmentation and object detection are often used in similar contexts.

CHAPTER 4

Building an Image Segmentation Model

Images around us come in different textures, patterns, shapes, and sizes. They carry an enormous amount of information which can easily be understood by the human eye and brain, but which are less easily understood by computers. Image segmentation is a problem set wherein we try to train computers to understand images so that they can separate dissimilar objects and group similar objects. This can be in the form of similar pixel intensities or similar textures and shapes.

There are a lot of algorithms that have been developed and have been used to segment images. Just like object detection separates objects, image segmentation identifies more similar objects from less similar objects. If we consider the concepts used in basic clustering methods such as k-means, we know how data points align themselves near similar data.

For example, let's assume there are two kinds of apples and two kinds of oranges placed in a bowl. If we look at the features, we can start by sorting the edible data points. We can separate the data points into two groups when we think about textures or color. When we add cost as another feature, we can separate the data points into four clusters. Likewise, this sequence can help us locate dissimilarities or similarities in data points and group by clusters. This is immensely helpful in the

© Akshay Kulkarni, Adarsha Shivananda, and Nitin Ranjan Sharma 2022
A. Kulkarni et al., *Computer Vision Projects with PyTorch*,
https://doi.org/10.1007/978-1-4842-8273-1_4

biomedical field and for self-driving cars. This is still an area of active research and is mostly coupled with object detection frameworks. We will go through the basics and then consider some examples. Let's get started.

Image Segmentation

The subjective nature of segmentation is based on the type of domain we are dealing with. There are two types of segmentation—semantic and instance. When we are working with semantic segmentation, the pixels from similar objects are considered to be one class, but there is no separation within the objects. If we imagine the scenario in real-time, when there is an image of multiple cars on a highway, the segmentation will group all the cars and separate those groups from the roadside or the scenery.

Let's consider an example. Figure 4-1a shows a highway with cars. Alongside multiple cars, there is grass and some trees on the side of the highway.

Figure 4-1a. *Raw input image*

Now consider a patch of pixels taken from the raw input and passed from a convolutional neural network that is capable of classifying the objects in the input (see Figure 4-1b). This will give us the output like the patch in the figure that belongs to a car. Next, we try to map the central

130

pixel to the car and iterate through the entire image like this. This will give us the segmentation separation (semantic) in the image. It will separate the cars from the trees and from the road. An important thing to note here is that all the cars will belong to the same class.

Figure 4-1b. *Taking a patch from input*

Another workaround for this type of problem is to run through a convolutional neural network classifier, which doesn't have any downsampling, and use that to classify each pixel and thus cluster similar objects together.

These all prove great until we want to distinguish between the classes. For example, say there are multiple cars and we want to classify each car separately. In this scenario, instance segmentation comes in, wherein every pixel is mapped to the specific class and objects are separated by labeling the pixels with the appropriate class. The idea of semantic segmentation goes back to non-learnable techniques used in image processing, whereas instance segmentation is a fairly new concept.

One of the basic approaches that came out when we started with instance segmentation was a near replica of the R-CNN approach. But we are predicting segments instead of regions.

Look at the process diagram in Figure 4-2. The image is passed onto a segment proposal network, which gives segments of the images. On the one hand, the segments can form a bounding box and be passed onto the box convolutional neural network to generate features. On the other hand, the segments are taken in and a background masking transformation is applied. It just takes in the mean values of the image and converts the background of the object to black. Once the segment is masked, it is passed to the region convolutional neural network for a different set of features.

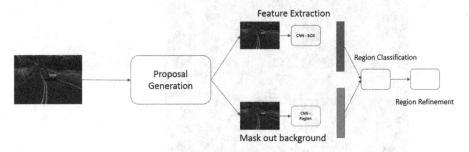

Figure 4-2. Instance segmentation process flow

What we have here is a combination of the box image and the region pulled by the network. These will be combined and then further classified based on the object instance they contain. A secondary step is also added beyond this when it refines the segmentation region.

These are simply experimental segmentation technique setups. Other methodical improvements have been made, including cascading networks, which are similar to Faster R-CNN, hypercolumns, etc.

These are multiple differences between semantic segmentation and instance segmentation.

With semantic segmentation:

- All the pixels are classified.

- Fully convolutional models are used.

- Downsampling is used in various approaches followed by learnable upsampling techniques to re-create the image.

- In case ResNet-like architectures are used, skip connections are used.

With instance segmentation:

- Not only every pixel is classified, but the instance is also detected.

- The process nearly follows the object detection architecture.

Pretrained Support from PyTorch

PyTorch has been growing at a much faster pace than any other framework. It has a plethora of modules and classes. Since it is close to Python, it is easier to adapt to the framework. There is a push toward selecting the PyTorch framework in deep learning in general and utilizing the vast resources to make impactful changes.

Like object detection, segmentation is also on the heavier side of architecture. It is not always easy or ideal to train these models from scratch. The training time is quite high in CPU, and even though GPU helps in this scenario, it's not by a lot. Due to all the training process constraints, we might opt for a transfer learning technique. This helps us leverage the rich information already extracted onto the work. The models are trained in diversified datasets and are generalized to handle variations in most of the problems we will encounter. Let's dive into some of the amazing models in the torch repository.

Semantic Segmentation

- **Fully convolutional neural network.** As proposed by the paper, a fully convolutional network is trained end-to-end on a semantic segmentation task. It's a convolutional neural network block followed by a pixel-wise prediction.

- **Use of atrous convolutions to semantic segmentation (DeepLavV3).** In the architecture, parallel stacks of atrous convolutions are used to capture the multi-scale context by altering receptive fields.

- **Lite Reduced Atrous Spatial Pyramid Pooling (LR-ASPP).** An advanced version of MobileNetV3, which is created with the help of NAS (Neural Architecture Search).

We will use one pretrained model to evaluate the model on our image. The model has a lot of parameters, so running inference will be slower on a CPU or low configuration infrastructure. If we are using Colab, we can switch on the GPU as infra support and run the code.

Let's start with the basic `imports` required for the configuration. The model is pretrained and placed in Torchvision. We will be importing the model as well.

```
import numpy as np

import torch

import matplotlib.pyplot as plt
## torchvision related imports
import torchvision.transforms.functional as F
from torchvision.io import read_image
from torchvision.utils import draw_bounding_boxes
from torchvision.utils import make_grid
## models and transforms
from torchvision.transforms.functional import convert_
image_dtype
from torchvision.models.segmentation import fcn_resnet50
```

So far we imported the Torch and Torchvision-related functions. We need to build all the utility functions that can be reused throughout the code. This is an effective way to refactor the code and remove unnecessary repetition. In this case, we need to display the images so we can use an image visualizer.

```
## utilities for multiple images
def img_show(images):

    if not isinstance(images, list):
        ## generalise cast images to list
        images = [images]
    fig, axis = plt.subplots(ncols=len(images), squeeze=False)
    for i, image in enumerate(images):
        image = image.detach() # detached from current DAG, no
        gradient
        image = F.to_pil_image(image)
        axis[0, i].imshow(np.asarray(image))
        axis[0, i].set(xticklabels=[], yticklabels=[],
        xticks=[], yticks=[])
```

This code accepts multiple images or a single image. It checks for the object to be a list, and if it is not, it converts that to a list. The axes are assigned according to an iterable of images. The images are detached from the DAG and the gradient is not calculated for these variables.

Following the utility function, let's get a sample image and configure it to work on the segmentation process.

```
## get an image on which segmentation needs to be done
img1 = read_image("/content/semantic_example_highway.jpg")

box_car = torch.tensor([ [170, 70, 220, 120]], dtype=torch.
float) ## (xmin,ymin,xmax,ymax)
colors = ["blue"]
```

```
check_box = draw_bounding_boxes(img1, box_car, colors=colors,
width=2)
img_show(check_box)
```

```
## batch for images
batch_imgs = torch.stack([img1])
batch_torch = convert_image_dtype(batch_imgs,
dtype=torch.float)
```

The image needs to be uploaded and placed in an accessible location. The image includes multiple cars, as of now, we are putting one box on top of it with (X_{min}, Y_{min}, X_{max}, and Y_{max}). These values need to be adjusted for the fully convolutional network to understand the presence of the object. Eventually, the batches of images are converted to tensors before being stacked for the model.

Now, let's load the model and get it ready for evaluation.

```
model = fcn_resnet50(pretrained=True, progress=False)
## switching on eval mode
model = model.eval()
# standard normalizing based on train config
normalized_batch_torch = F.normalize(batch_torch, mean=(0.485,
0.456, 0.406), std=(0.229, 0.224, 0.225))
result = model(normalized_batch_torch)['out']
```

As discussed earlier, fcn_resnet50 is downloaded from the repository, which is already trained. The model is set to the eval step. The batch, which was created in the earlier step, is now normalized based on the trained model configuration.

It's now time to pass our image through the model.

```
classes = [
    '__background__', 'aeroplane', 'bicycle', 'bird', 'boat',
    'bottle', 'bus',
```

```
    'car', 'cat', 'chair', 'cow', 'diningtable', 'dog',
    'horse', 'motorbike',
    'person', 'pottedplant', 'sheep', 'sofa', 'train',
    'tvmonitor'
]
class_to_idx = {cls: idx for (idx, cls) in enumerate(classes)}

normalized_out_masks = torch.nn.functional.
softmax(result, dim=1)

car_mask = [
    normalized_out_masks[img_idx, class_to_idx[cls]]
    for img_idx in range(batch_torch.shape[0])
    for cls in ('car', 'pottedplant','bus')
]

img_show(car_mask)
```

We are defining the list with all the probable classes and getting the batch image results to pass through the softmax layer. The masks are then plotted. This example gives us the process flow for the semantic segmentation we discussed earlier. It shows how we can get any data and prepare it according to the model. We load a model and run inference on the model to get the masks.

Instance Segmentation

We were working on semantic segmentation to generate masks of objects and then superimpose them on the original images. But, what about instance segmentation? We will now look at some of the pretrained models that we can leverage for generating masks.

Models for detection and masking:

- **Faster R-CNN.** This research introduced a region proposal network that simultaneously predicts the object bounding box and the objectness core corresponding to the bounding box. It solves the bottleneck related to the earlier papers.

- **Mask R-CNN.** This process extends Faster R-CNN and employs object detection and generating masking on the image.

- **RetinaNet.** This paper did some amazing improvements to the two-stage detectors in terms of accuracy and speed. It takes care of all these using the new concept of *focal loss*.

- **Single Shot Detector.** The paper explains that a objectness score is generated for the default bounding box and is refined based on the object.

These models are mainly trained on COCO datasets and are capable of handling predictions.

For faster R-CNN:

```
x = [torch.rand(3, 300, 400), torch.rand(3, 500, 400)]
```

```
faster_rcnn_model = torchvision.models.detection.fasterrcnn_
resnet50_fpn(pretrained=True)
faster_rcnn_model.eval()
result = faster_rcnn_model (x)
```

For MobileNet:

```
x = [torch.rand(3, 300, 400), torch.rand(3, 500, 400)]
```

```
mobilenet_model = torchvision.models.detection.fasterrcnn_
mobilenet_v3_large_fpn(pretrained=True)
mobilenet_model.eval()
result = mobilenet_model(x)
```

For RetinaNet :

```
x = [torch.rand(3, 300, 400), torch.rand(3, 500, 400)]
```

```
retinanet_model = torchvision.models.detection.retinanet_
resnet50_fpn(pretrained=True)
retinanet_model.eval()
result = retinanet_model(x)
```

For Single Shot Detection :

```
x = [torch.rand(3, 300, 400), torch.rand(3, 500, 400)]
```

```
ssd_model = torchvision.models.detection.ssd300_
vgg16(pretrained=True)
ssd_model.eval()
result = ssd_model(x)
```

For all these instances, we are extracting the model from the PyTorch repository and using it to run inference.

Fine-Tuning the Model

Using a pretrained model can get predictions if we are working in a domain that has classes that have been labeled and trained by the researchers. If it's closely related but we don't have the exact classes, we can expect variations. This is one of the most important reasons to train and classify for our project.

In this section, we explore the code and explain in detail the necessary steps for fine-tuning an existing model to enhance the predictive power of the model to suit our purposes. As we have established, segmentation includes additional capabilities to the ones that can identify classes in an image.

The problem set that we will be using is an open source dataset found at https://www.cis.upenn.edu/~jshi/ped_html/PennFudanPed.zip.

This dataset contains pedestrian data, which we will be fine-tuning. Various use cases use segmentation as an output to determine the decision steps. Identifying pedestrians is an important use case for self-driving cars when making the decision as to which direction they need to move in a split second. For these reasons, accuracies also need to be high enough for these models.

Let's look through the basic imports for the process flow.

The project setup stands to be one of the most decisive parts of any project. In this project, we can use a Jupyter notebook instance or a Colab notebook for training purposes.

First, we download the dataset from the source using the wget command.

```
## extracting traffic data
!wget https://www.cis.upenn.edu/~jshi/ped_html/
PennFudanPed.zip .
!unzip PennFudanPed.zip
```

The exclamation point (!) helps the Colab cells identify those as shell scripts. Once downloaded, the unzip command unzips the package. One thing to notice is that the commands are Linux based and the backend OS is also assumed to be Linux.

Once we have the dataset in our system, we can run the basic imports that are required for the project.

```
## basic imports
import os
import numpy as np

## torch imports
import torch
import torch.utils.data
from torch.utils.data import Dataset

## torchvision imports
import torchvision
import torchvision.transforms as T
from torchvision.models.detection.faster_rcnn import
FastRCNNPredictor
from torchvision.models.detection.mask_rcnn import
MaskRCNNPredictor

## image utilities
from PIL import Image
import matplotlib.pyplot as plt

## code utilities
import random
import cv2
```

Note that we are importing Torch and Torchvision-related packages.
We are also getting MaskRCNN and FasterRCNN pretrained models. Let's
now import the PyTorch basic training framework. This will help us extend
the functions and avoid rewriting complex code.

```
## clone the PyTorch repository to setup exact directory
structures as the original trained
!git clone https://github.com/pytorch/vision.git
%cd vision
!git checkout v0.3.0
```

```
!cp references/detection/engine.py ../
!cp references/detection/transforms.py ../
!cp references/detection/utils.py ../
!cp references/detection/coco_utils.py ../
!cp references/detection/coco_eval.py ../
```

Once we have the basic framework ready, we will be copying the important Python scripts that we will be using, such as the engine, transforms, utils, coco_utils, and coco_eval.

Once these imports are done and we have verified that the files are on the same infrastructure where we are running the code, we can run a few more imports.

```
## imports from the PyTorch repo
import utils
import transforms as T
from engine import train_one_epoch, evaluate
```

These imports are based on the code in the PyTorch training framework and scripts. Once this is done, let's look at the creation of the custom dataset class required to fine-tune the models.

```
class CustomDataset(Dataset):
    def __init__(self, dir_path, transforms=None):
        ## initializing object attributes
        self.transforms = transforms
        self.dir_path = dir_path
        ## from dir_path
        ## added list of masks from the PedMasks directory
        self.mask_list = list(sorted(os.listdir(os.path.
        join(dir_path, "PedMasks"))))
        ## added list of actual images from directory lists
```

```python
        self.image_list = list(sorted(os.listdir(os.path.
        join(dir_path, "PNGImages"))))

    def __getitem__(self, idx):
        # get images and mask
        img_path = os.path.join(self.dir_path, "PNGImages",
        self.image_list[idx])
        mask_path = os.path.join(self.dir_path, "PedMasks",
        self.mask_list[idx])
        image_obj = Image.open(img_path).convert("RGB")
        mask_obj = Image.open(mask_path)

        mask_obj = np.array(mask_obj)
        obj_ids = np.unique(mask_obj)
        # background has the first id so excluding that
        obj_ids = obj_ids[1:]
        # splitting mask into binaries
        masks_obj = mask_obj == obj_ids[:, None, None]
        # bounding box
        num_objs = len(obj_ids)
        bboxes = []
        for i in range(num_objs):
            pos = np.where(masks_obj[i])

            xmax = np.max(pos[1])
            xmin = np.min(pos[1])
            ymax = np.max(pos[0])
            ymin = np.min(pos[0])

            bboxes.append([xmin, ymin, xmax, ymax])

        image_id = torch.tensor([idx])
        masks_obj = torch.as_tensor(masks_obj,
        dtype=torch.uint8)
```

```
    bboxes = torch.as_tensor(bboxes, dtype=torch.float32)
    labels = torch.ones((num_objs,), dtype=torch.int64)

    area = (bboxes[:, 3] - bboxes[:, 1]) * (bboxes[:, 2] -
    bboxes[:, 0])

    iscrowd = torch.zeros((num_objs,), dtype=torch.int64)

    target = {}
    target["image_id"] = image_id
    target["masks"] = masks_obj
    target["boxes"] = bboxes
    target["labels"] = labels

    target["area"] = area
    target["iscrowd"] = iscrowd

    if self.transforms is not None:
        image_obj, target = self.transforms(image_
        obj, target)

    return image_obj, target

def __len__(self):
    return len(self.image_list)
```

Custom dataset creation is one standard technique to incorporate the newer dataset into the training pipeline. The important points to look for in the code are as follows:

- We are extending the Dataset class from PyTorch.

- We are defining three important functions to the class—initialization, get_item, and len.

- We are initializing transforms, which can be different for testing, validation, and training.

- We are defining the bounding boxes.

- We are defining the target.

Once we are done with the dataset creation, we have to modify the model according to the new data. Let's look at the code for that.

```
def modify_model(classes_num):
    # model already trained on COCO loaded from PyTorch
    repository
    maskrcnn_model = torchvision.models.detection.maskrcnn_
    resnet50_fpn(pretrained=True)

    # number of input features identification
    in_features = maskrcnn_model.roi_heads.box_predictor.cls_
    score.in_features

    # head is changed
    maskrcnn_model.roi_heads.box_predictor =
    FastRCNNPredictor(in_features, classes_num)

    in_features_mask = maskrcnn_model.roi_heads.mask_predictor.
    conv5_mask.in_channels
    hidden_layer = 256

    maskrcnn_model.roi_heads.mask_predictor =
    MaskRCNNPredictor(in_features_mask, hidden_layer,
    num_classes)

    return maskrcnn_model
```

These steps change the head configuration of the model. After this, we will try to get transformations on the data. This will prepare the data for training.

```
def get_transform_data(train):
    transforms = []
```

145

```python
        # PIL image to tensor for PyTorch model
        transforms.append(T.ToTensor())
        if train:
            # basic image augmentation techniques
            ## can add few more for experimentation
            transforms.append(T.RandomHorizontalFlip(0.5))
        return T.Compose(transforms)

# get the traffic data to transform
train_dataset = CustomDataset('/content/PennFudanPed', get_
transform_data(train=True))
test_dataset = CustomDataset('/content/PennFudanPed', get_
transform_data(train=False))

# train test split
torch.manual_seed(1)
indices = torch.randperm(len(train_dataset)).tolist()
train_dataset = torch.utils.data.Subset(train_dataset,
indices[:-50])
test_dataset = torch.utils.data.Subset(test_dataset,
indices[-50:])

# define training and validation data loaders
train_data_loader = torch.utils.data.DataLoader(
    train_dataset, batch_size=2, shuffle=True, num_workers=4,
    collate_fn=utils.collate_fn)

test_data_loader = torch.utils.data.DataLoader(
    test_dataset, batch_size=1, shuffle=False, num_workers=4,
    collate_fn=utils.collate_fn)
```

The important points to look for in the transform data are:

- Converting the data into tensors so that it can be used in the PyTorch DAG.

- For training, we are using transformation or augmentation techniques. For all other purposes such as the test and validation steps, we will not be using any augmentation techniques.

- We create the train and the test dataset from the custom dataset class we built in an earlier stage.

- Once the custom data is defined, we will use that to create an iterable, which will directly help us with the training.

- The iterable is also called a *data loader*.

Once the data loader is created, we can move to the training part. We will be defining the device on which training needs to take place. We will define the optimizer and the learning rate scheduler.

```
device = torch.device('cuda') if torch.cuda.is_available() else
torch.device('cpu')
# since we are dealing with persons and background number of
classes become 2
num_classes = 2

final_model = modify_model(num_classes)
# model to GPU or CPU if GPU not available
final_model.to(device)

## getting SGD optimizer
params = [p for p in final_model.parameters() if
p.requires_grad]
optimizer = torch.optim.SGD(params,
```

```
                                     lr=0.005,
                                     momentum=0.9,
                                     weight_decay=0.0005)

# setting up the step learning rate

lr_scheduler = torch.optim.lr_scheduler.StepLR(optimizer,
                                               step_size=2,
                                               gamma=0.1)
```

The key points to note from the code are as follows:

- The device where the model is residing should be the same as the data it is being trained with. There should not be any cross-infrastructure transitions between models and data. If the data is too large to fit on one GPU, batches of data should reside in the same system as the model.

- Setting up optimizers and the learning rate scheduler.

- It should be noted that, when we are training complex networks, a fixed learning rate would not help us train faster or more efficiently. Learning rate alone is one of the most important hyperparameters for our training process and we need to handle it with care.

Now that we have set our model parameters, let's run through a few epochs to check the training process.

```
# setting up epochs
num_epochs = 5

for epoch in range(num_epochs):
    ## using train_one_epoch from pytorch helper
    function itself
```

```
# getting used to fine tuning framework
train_one_epoch(final_model, optimizer, train_data_loader,
device, epoch, print_freq=10)
# updating the weights and learning rates
lr_scheduler.step()
# get the evaluation result from the resultant change
in weights
evaluate(final_model, test_data_loader, device=device)
```

For the sake of simplicity, we ran the model through five epochs, but for a better result, you should run it longer. One of the most important aspects of the training process is to check the generated logs. They should give us a fair understanding of how the data is run through the model and how the training process is established. Let's quickly glance through the logs generated by the epochs.

```
Epoch: [0]  [ 0/60]  eta: 0:02:17  lr: 0.000090  loss: 2.7890
(2.7890)  loss_classifier: 0.7472 (0.7472)  loss_box_reg:
0.3405 (0.3405)  loss_mask: 1.6637 (1.6637)  loss_objectness:
0.0351 (0.0351)  loss_rpn_box_reg: 0.0025 (0.0025)  time:
2.2894  data: 0.4357  max mem: 2161
Epoch: [0]  [10/60]  eta: 0:01:26  lr: 0.000936  loss: 1.3992
(1.7301)  loss_classifier: 0.5175 (0.4831)  loss_box_reg:
0.2951 (0.2971)  loss_mask: 0.7160 (0.9201)  loss_objectness:
0.0279 (0.0249)  loss_rpn_box_reg: 0.0045 (0.0048)  time:
1.7208  data: 0.0469  max mem: 3316
Epoch: [0]  [20/60]  eta: 0:01:05  lr: 0.001783  loss: 1.0006
(1.2323)  loss_classifier: 0.2196 (0.3358)  loss_box_reg:
0.2905 (0.2854)  loss_mask: 0.3228 (0.5877)  loss_objectness:
0.0172 (0.0188)  loss_rpn_box_reg: 0.0042 (0.0045)  time:
1.6055  data: 0.0096  max mem: 3316
```

Epoch: [0] [30/60] eta: 0:00:49 lr: 0.002629 loss: 0.5668
(1.0164) loss_classifier: 0.0936 (0.2558) loss_box_reg:
0.2643 (0.2860) loss_mask: 0.1797 (0.4540) loss_objectness:
0.0056 (0.0156) loss_rpn_box_reg: 0.0045 (0.0050) time:
1.6322 data: 0.0108 max mem: 3316
Epoch: [0] [40/60] eta: 0:00:33 lr: 0.003476 loss: 0.4461
(0.8835) loss_classifier: 0.0639 (0.2070) loss_box_reg:
0.2200 (0.2681) loss_mask: 0.1693 (0.3904) loss_objectness:
0.0028 (0.0126) loss_rpn_box_reg: 0.0057 (0.0054) time:
1.6640 data: 0.0107 max mem: 3316
Epoch: [0] [50/60] eta: 0:00:16 lr: 0.004323 loss: 0.3779
(0.7842) loss_classifier: 0.0396 (0.1749) loss_box_reg:
0.1619 (0.2452) loss_mask: 0.1670 (0.3476) loss_objectness:
0.0014 (0.0107) loss_rpn_box_reg: 0.0051 (0.0058) time:
1.5650 data: 0.0107 max mem: 3316
Epoch: [0] [59/60] eta: 0:00:01 lr: 0.005000 loss: 0.3066
(0.7143) loss_classifier: 0.0329 (0.1549) loss_box_reg:
0.1074 (0.2265) loss_mask: 0.1508 (0.3172) loss_objectness:
0.0022 (0.0097) loss_rpn_box_reg: 0.0052 (0.0059) time:
1.5627 data: 0.0109 max mem: 3316
Epoch: [0] Total time: 0:01:37 (1.6202 s / it)
creating index...
index created!
Test: [0/50] eta: 0:00:27 model_time: 0.3958
(0.3958) evaluator_time: 0.0052 (0.0052) time: 0.5474 data:
0.1449 max mem: 3316
Test: [49/50] eta: 0:00:00 model_time: 0.3451
(0.3489) evaluator_time: 0.0061 (0.0110) time: 0.3666 data:
0.0055 max mem: 3316
Test: Total time: 0:00:18 (0.3715 s / it)

Averaged stats: model_time: 0.3451 (0.3489) evaluator_time:
0.0061 (0.0110)
Accumulating evaluation results...
DONE (t=0.01s).
Accumulating evaluation results...
DONE (t=0.01s).
IoU metric: bbox
 Average Precision (AP) @[IoU=0.50:0.95 | area=   all |
 maxDets=100] = 0.690
 Average Precision (AP) @[IoU=0.50 | area=   all |
 maxDets=100] = 0.976
 Average Precision (AP) @[IoU=0.75 | area=   all |
 maxDets=100] = 0.863
 Average Precision (AP) @[IoU=0.50:0.95 | area= small |
 maxDets=100] = -1.000
 Average Precision (AP) @[IoU=0.50:0.95 | area=medium |
 maxDets=100] = 0.363
 Average Precision (AP) @[IoU=0.50:0.95 | area= large |
 maxDets=100] = 0.708
 Average Recall   (AR) @[IoU=0.50:0.95 | area=   all |
 maxDets= 1] = 0.311
 Average Recall   (AR) @[IoU=0.50:0.95 | area=   all |
 maxDets= 10] = 0.747
 Average Recall   (AR) @[IoU=0.50:0.95 | area=   all |
 maxDets=100] = 0.747
 Average Recall   (AR) @[IoU=0.50:0.95 | area= small |
 maxDets=100] = -1.000
 Average Recall   (AR) @[IoU=0.50:0.95 | area=medium |
 maxDets=100] = 0.637
 Average Recall   (AR) @[IoU=0.50:0.95 | area= large |
 maxDets=100] = 0.755

```
IoU metric: segm
 Average Precision  (AP) @[ IoU=0.50:0.95 | area=   all |
 maxDets=100 ] = 0.722
 Average Precision  (AP) @[ IoU=0.50      | area=   all |
 maxDets=100 ] = 0.976
 Average Precision  (AP) @[ IoU=0.75      | area=   all |
 maxDets=100 ] = 0.886
 Average Precision  (AP) @[ IoU=0.50:0.95 | area= small |
 maxDets=100 ] = -1.000
 Average Precision  (AP) @[ IoU=0.50:0.95 | area=medium |
 maxDets=100 ] = 0.448
 Average Precision  (AP) @[ IoU=0.50:0.95 | area= large |
 maxDets=100 ] = 0.740
 Average Recall     (AR) @[ IoU=0.50:0.95 | area=   all |
 maxDets=  1 ] = 0.325
 Average Recall     (AR) @[ IoU=0.50:0.95 | area=   all |
 maxDets= 10 ] = 0.760
 Average Recall     (AR) @[ IoU=0.50:0.95 | area=   all |
 maxDets=100 ] = 0.761
 Average Recall     (AR) @[ IoU=0.50:0.95 | area= small |
 maxDets=100 ] = -1.000
 Average Recall     (AR) @[ IoU=0.50:0.95 | area=medium |
 maxDets=100 ] = 0.675
 Average Recall     (AR) @[ IoU=0.50:0.95 | area= large |
 maxDets=100 ] = 0.767
```

Here are some essential points that need to be noted while checking on the logs for any training network:

- Since we have used a stepwise learning rate change, it can be easily noticed in the log verbosity.

- The loss from the classifier and the loss from objectness can be noted for each batch.

- Other important aspects that we should be checking are the average precision and average recall.

- ETA and memory allocation can help us estimate the computation for a larger evaluation of the model.

Now that we have trained the model, we can opt to save the model by using the torch save command. We can also use the Save Dictionary option, which has more advantages rather than just saving the pickled form. When we are saving a dictionary, we can essentially alter the dictionary when we want, but that might not be the case when we are storing it in the form of a pickle. The pickle stores the directory paths and the model parameters and is very difficult to change or alter.

```
## saving the model full version
## can opt for state dict version of saving
torch.save(final_model, 'mask-rcnn-fine_tuned.pt')
```

Since we have now a trained model with us, we can start our inference. It will start with the eval mode, which switches the model to evaluation mode. No gradients are computed.

```
# pytorch help to set the model to eval mode
final_model.eval()
CLASSES = ['__background__', 'pedestrian']
device = torch.device('cuda') if torch.cuda.is_available() else
torch.device('cpu')
final_model.to(device)
```

This will also help us describe the model that is going to be used for inference.

```
MaskRCNN(
  (transform): GeneralizedRCNNTransform(
    Normalize(mean=[0.485, 0.456, 0.406], std=[0.229,
    0.224, 0.225])
    Resize(min_size=(800,), max_size=1333, mode='bilinear')
  )
  (backbone): BackboneWithFPN(
    (body): IntermediateLayerGetter(
      (conv1): Conv2d(3, 64, kernel_size=(7, 7), stride=(2, 2),
      padding=(3, 3), bias=False)
      (bn1): FrozenBatchNorm2d(64, eps=0.0)
      (relu): ReLU(inplace=True)
      (maxpool): MaxPool2d(kernel_size=3, stride=2, padding=1,
      dilation=1, ceil_mode=False)
      (layer1): Sequential(
        (0): Bottleneck(
          (conv1): Conv2d(64, 64, kernel_size=(1, 1),
          stride=(1, 1), bias=False)
          (bn1): FrozenBatchNorm2d(64, eps=0.0)
          (conv2): Conv2d(64, 64, kernel_size=(3, 3),
          stride=(1, 1), padding=(1, 1), bias=False)
          (bn2): FrozenBatchNorm2d(64, eps=0.0)
          (conv3): Conv2d(64, 256, kernel_size=(1, 1),
          stride=(1, 1), bias=False)
          (bn3): FrozenBatchNorm2d(256, eps=0.0)
          (relu): ReLU(inplace=True)
          (downsample): Sequential(
            (0): Conv2d(64, 256, kernel_size=(1, 1),
            stride=(1, 1), bias=False)
            (1): FrozenBatchNorm2d(256, eps=0.0)
          )
        )
```

```
(1): Bottleneck(
  (conv1): Conv2d(256, 64, kernel_size=(1, 1),
  stride=(1, 1), bias=False)
  (bn1): FrozenBatchNorm2d(64, eps=0.0)
  (conv2): Conv2d(64, 64, kernel_size=(3, 3),
  stride=(1, 1), padding=(1, 1), bias=False)
  (bn2): FrozenBatchNorm2d(64, eps=0.0)
  (conv3): Conv2d(64, 256, kernel_size=(1, 1),
  stride=(1, 1), bias=False)
  (bn3): FrozenBatchNorm2d(256, eps=0.0)
  (relu): ReLU(inplace=True)
)
(2): Bottleneck(
  (conv1): Conv2d(256, 64, kernel_size=(1, 1),
  stride=(1, 1), bias=False)
  (bn1): FrozenBatchNorm2d(64, eps=0.0)
  (conv2): Conv2d(64, 64, kernel_size=(3, 3),
  stride=(1, 1), padding=(1, 1), bias=False)
  (bn2): FrozenBatchNorm2d(64, eps=0.0)
  (conv3): Conv2d(64, 256, kernel_size=(1, 1),
  stride=(1, 1), bias=False)
  (bn3): FrozenBatchNorm2d(256, eps=0.0)
  (relu): ReLU(inplace=True)
)
)
(layer2): Sequential(
  (0): Bottleneck(
    (conv1): Conv2d(256, 128, kernel_size=(1, 1),
    stride=(1, 1), bias=False)
    (bn1): FrozenBatchNorm2d(128, eps=0.0)
```

```
(conv2): Conv2d(128, 128, kernel_size=(3, 3),
stride=(2, 2), padding=(1, 1), bias=False)
(bn2): FrozenBatchNorm2d(128, eps=0.0)
(conv3): Conv2d(128, 512, kernel_size=(1, 1),
stride=(1, 1), bias=False)
(bn3): FrozenBatchNorm2d(512, eps=0.0)
(relu): ReLU(inplace=True)
(downsample): Sequential(
  (0): Conv2d(256, 512, kernel_size=(1, 1),
  stride=(2, 2), bias=False)
  (1): FrozenBatchNorm2d(512, eps=0.0)
)
)

(1): Bottleneck(
(conv1): Conv2d(512, 128, kernel_size=(1, 1),
stride=(1, 1), bias=False)
(bn1): FrozenBatchNorm2d(128, eps=0.0)
(conv2): Conv2d(128, 128, kernel_size=(3, 3),
stride=(1, 1), padding=(1, 1), bias=False)
(bn2): FrozenBatchNorm2d(128, eps=0.0)
(conv3): Conv2d(128, 512, kernel_size=(1, 1),
stride=(1, 1), bias=False)
(bn3): FrozenBatchNorm2d(512, eps=0.0)
(relu): ReLU(inplace=True)
)

(2): Bottleneck(
(conv1): Conv2d(512, 128, kernel_size=(1, 1),
stride=(1, 1), bias=False)
(bn1): FrozenBatchNorm2d(128, eps=0.0)
(conv2): Conv2d(128, 128, kernel_size=(3, 3),
stride=(1, 1), padding=(1, 1), bias=False)
```

```
      (bn2): FrozenBatchNorm2d(128, eps=0.0)
      (conv3): Conv2d(128, 512, kernel_size=(1, 1),
      stride=(1, 1), bias=False)
      (bn3): FrozenBatchNorm2d(512, eps=0.0)
      (relu): ReLU(inplace=True)
    )
    (3): Bottleneck(
      (conv1): Conv2d(512, 128, kernel_size=(1, 1),
      stride=(1, 1), bias=False)
      (bn1): FrozenBatchNorm2d(128, eps=0.0)
      (conv2): Conv2d(128, 128, kernel_size=(3, 3),
      stride=(1, 1), padding=(1, 1), bias=False)
      (bn2): FrozenBatchNorm2d(128, eps=0.0)
      (conv3): Conv2d(128, 512, kernel_size=(1, 1),
      stride=(1, 1), bias=False)
      (bn3): FrozenBatchNorm2d(512, eps=0.0)
      (relu): ReLU(inplace=True)
    )
  )
  (layer3): Sequential(
    (0): Bottleneck(
      (conv1): Conv2d(512, 256, kernel_size=(1, 1),
      stride=(1, 1), bias=False)
      (bn1): FrozenBatchNorm2d(256, eps=0.0)
      (conv2): Conv2d(256, 256, kernel_size=(3, 3),
      stride=(2, 2), padding=(1, 1), bias=False)
      (bn2): FrozenBatchNorm2d(256, eps=0.0)
      (conv3): Conv2d(256, 1024, kernel_size=(1, 1),
      stride=(1, 1), bias=False)
      (bn3): FrozenBatchNorm2d(1024, eps=0.0)
      (relu): ReLU(inplace=True)
```

```
      (downsample): Sequential(
        (0): Conv2d(512, 1024, kernel_size=(1, 1),
        stride=(2, 2), bias=False)
        (1): FrozenBatchNorm2d(1024, eps=0.0)
      )
    )
    (1): Bottleneck(
      (conv1): Conv2d(1024, 256, kernel_size=(1, 1),
      stride=(1, 1), bias=False)
      (bn1): FrozenBatchNorm2d(256, eps=0.0)
      (conv2): Conv2d(256, 256, kernel_size=(3, 3),
      stride=(1, 1), padding=(1, 1), bias=False)
      (bn2): FrozenBatchNorm2d(256, eps=0.0)
      (conv3): Conv2d(256, 1024, kernel_size=(1, 1),
      stride=(1, 1), bias=False)
      (bn3): FrozenBatchNorm2d(1024, eps=0.0)
      (relu): ReLU(inplace=True)
    )
    (2): Bottleneck(
      (conv1): Conv2d(1024, 256, kernel_size=(1, 1),
      stride=(1, 1), bias=False)
      (bn1): FrozenBatchNorm2d(256, eps=0.0)
      (conv2): Conv2d(256, 256, kernel_size=(3, 3),
      stride=(1, 1), padding=(1, 1), bias=False)
      (bn2): FrozenBatchNorm2d(256, eps=0.0)
      (conv3): Conv2d(256, 1024, kernel_size=(1, 1),
      stride=(1, 1), bias=False)
      (bn3): FrozenBatchNorm2d(1024, eps=0.0)
      (relu): ReLU(inplace=True)
    )
```

```
(3): Bottleneck(
  (conv1): Conv2d(1024, 256, kernel_size=(1, 1),
  stride=(1, 1), bias=False)
  (bn1): FrozenBatchNorm2d(256, eps=0.0)
  (conv2): Conv2d(256, 256, kernel_size=(3, 3),
  stride=(1, 1), padding=(1, 1), bias=False)
  (bn2): FrozenBatchNorm2d(256, eps=0.0)
  (conv3): Conv2d(256, 1024, kernel_size=(1, 1),
  stride=(1, 1), bias=False)
  (bn3): FrozenBatchNorm2d(1024, eps=0.0)
  (relu): ReLU(inplace=True)
)
(4): Bottleneck(
  (conv1): Conv2d(1024, 256, kernel_size=(1, 1),
  stride=(1, 1), bias=False)
  (bn1): FrozenBatchNorm2d(256, eps=0.0)
  (conv2): Conv2d(256, 256, kernel_size=(3, 3),
  stride=(1, 1), padding=(1, 1), bias=False)
  (bn2): FrozenBatchNorm2d(256, eps=0.0)
  (conv3): Conv2d(256, 1024, kernel_size=(1, 1),
  stride=(1, 1), bias=False)
  (bn3): FrozenBatchNorm2d(1024, eps=0.0)
  (relu): ReLU(inplace=True)
)
(5): Bottleneck(
  (conv1): Conv2d(1024, 256, kernel_size=(1, 1),
  stride=(1, 1), bias=False)
  (bn1): FrozenBatchNorm2d(256, eps=0.0)
  (conv2): Conv2d(256, 256, kernel_size=(3, 3),
  stride=(1, 1), padding=(1, 1), bias=False)
  (bn2): FrozenBatchNorm2d(256, eps=0.0)
```

```
            (conv3): Conv2d(256, 1024, kernel_size=(1, 1),
            stride=(1, 1), bias=False)
            (bn3): FrozenBatchNorm2d(1024, eps=0.0)
            (relu): ReLU(inplace=True)
          )
        )
        (layer4): Sequential(
          (0): Bottleneck(
            (conv1): Conv2d(1024, 512, kernel_size=(1, 1),
            stride=(1, 1), bias=False)
            (bn1): FrozenBatchNorm2d(512, eps=0.0)
            (conv2): Conv2d(512, 512, kernel_size=(3, 3),
            stride=(2, 2), padding=(1, 1), bias=False)
            (bn2): FrozenBatchNorm2d(512, eps=0.0)
            (conv3): Conv2d(512, 2048, kernel_size=(1, 1),
            stride=(1, 1), bias=False)
            (bn3): FrozenBatchNorm2d(2048, eps=0.0)
            (relu): ReLU(inplace=True)
            (downsample): Sequential(
              (0): Conv2d(1024, 2048, kernel_size=(1, 1),
              stride=(2, 2), bias=False)
              (1): FrozenBatchNorm2d(2048, eps=0.0)
            )
          )
          (1): Bottleneck(
            (conv1): Conv2d(2048, 512, kernel_size=(1, 1),
            stride=(1, 1), bias=False)
            (bn1): FrozenBatchNorm2d(512, eps=0.0)
            (conv2): Conv2d(512, 512, kernel_size=(3, 3),
            stride=(1, 1), padding=(1, 1), bias=False)
            (bn2): FrozenBatchNorm2d(512, eps=0.0)
```

```
      (conv3): Conv2d(512, 2048, kernel_size=(1, 1),
      stride=(1, 1), bias=False)
      (bn3): FrozenBatchNorm2d(2048, eps=0.0)
      (relu): ReLU(inplace=True)
    )
    (2): Bottleneck(
      (conv1): Conv2d(2048, 512, kernel_size=(1, 1),
      stride=(1, 1), bias=False)
      (bn1): FrozenBatchNorm2d(512, eps=0.0)
      (conv2): Conv2d(512, 512, kernel_size=(3, 3),
      stride=(1, 1), padding=(1, 1), bias=False)
      (bn2): FrozenBatchNorm2d(512, eps=0.0)
      (conv3): Conv2d(512, 2048, kernel_size=(1, 1),
      stride=(1, 1), bias=False)
      (bn3): FrozenBatchNorm2d(2048, eps=0.0)
      (relu): ReLU(inplace=True)
    )
  )
)
(fpn): FeaturePyramidNetwork(
  (inner_blocks): ModuleList(
    (0): Conv2d(256, 256, kernel_size=(1, 1),
    stride=(1, 1))
    (1): Conv2d(512, 256, kernel_size=(1, 1),
    stride=(1, 1))
    (2): Conv2d(1024, 256, kernel_size=(1, 1),
    stride=(1, 1))
    (3): Conv2d(2048, 256, kernel_size=(1, 1),
    stride=(1, 1))
  )
  (layer_blocks): ModuleList(
```

```
      (0): Conv2d(256, 256, kernel_size=(3, 3), stride=(1, 1),
      padding=(1, 1))
      (1): Conv2d(256, 256, kernel_size=(3, 3), stride=(1, 1),
      padding=(1, 1))
      (2): Conv2d(256, 256, kernel_size=(3, 3), stride=(1, 1),
      padding=(1, 1))
      (3): Conv2d(256, 256, kernel_size=(3, 3), stride=(1, 1),
      padding=(1, 1))
    )
    (extra_blocks): LastLevelMaxPool()
  )
)
(rpn): RegionProposalNetwork(
  (anchor_generator): AnchorGenerator()
  (head): RPNHead(
    (conv): Conv2d(256, 256, kernel_size=(3, 3), stride=(1, 1),
    padding=(1, 1))
    (cls_logits): Conv2d(256, 3, kernel_size=(1, 1),
    stride=(1, 1))
    (bbox_pred): Conv2d(256, 12, kernel_size=(1, 1),
    stride=(1, 1))
  )
)
(roi_heads): RoIHeads(
  (box_roi_pool): MultiScaleRoIAlign(featmap_names=['0', '1',
  '2', '3'], output_size=(7, 7), sampling_ratio=2)
  (box_head): TwoMLPHead(
    (fc6): Linear(in_features=12544, out_features=1024,
    bias=True)
    (fc7): Linear(in_features=1024, out_features=1024,
    bias=True)
  )
```

```
(box_predictor): FastRCNNPredictor(
  (cls_score): Linear(in_features=1024, out_features=2,
  bias=True)
  (bbox_pred): Linear(in_features=1024, out_features=8,
  bias=True)
)
(mask_roi_pool): MultiScaleRoIAlign(featmap_names=['0',
'1', '2', '3'], output_size=(14, 14), sampling_ratio=2)
(mask_head): MaskRCNNHeads(
  (mask_fcn1): Conv2d(256, 256, kernel_size=(3, 3),
  stride=(1, 1), padding=(1, 1))
  (relu1): ReLU(inplace=True)
  (mask_fcn2): Conv2d(256, 256, kernel_size=(3, 3),
  stride=(1, 1), padding=(1, 1))
  (relu2): ReLU(inplace=True)
  (mask_fcn3): Conv2d(256, 256, kernel_size=(3, 3),
  stride=(1, 1), padding=(1, 1))
  (relu3): ReLU(inplace=True)
  (mask_fcn4): Conv2d(256, 256, kernel_size=(3, 3),
  stride=(1, 1), padding=(1, 1))
  (relu4): ReLU(inplace=True)
)
(mask_predictor): MaskRCNNPredictor(
  (conv5_mask): ConvTranspose2d(256, 256, kernel_size=(2,
  2), stride=(2, 2))
  (relu): ReLU(inplace=True)
  (mask_fcn_logits): Conv2d(256, 2, kernel_size=(1, 1),
  stride=(1, 1))
)
)
)
```

This model definition can help us establish the architecture we are currently using and the alteration we can do around this. Once this is done let's write a few more lines of code to display the masks.

```
def get_mask_color(mask_conf):
    ## helper function to generate mask
    colour_option = [[0, 250, 0],[0, 0, 250],[250, 0, 0],
    [0, 250, 250],[250, 250, 0],[250, 0, 250],[75, 65, 170],
    [230, 75, 180],[235, 130, 40],[60, 140, 240],[40,
    180, 180]]
    blue = np.zeros_like(mask_conf).astype(np.uint8)
    green = np.zeros_like(mask_conf).astype(np.uint8)
    red = np.zeros_like(mask_conf).astype(np.uint8)

    red[mask_conf == 1], green[mask_conf == 1], blue[mask_conf
    == 1] = colour_option[random.randrange(0,10)]
    mask_color = np.stack([red, green, blue], axis=2)
    return mask_color

def generate_prediction(image_path, conf):
    ## helper function to generate predictions
    image = Image.open(image_path)
    transform = T.Compose([T.ToTensor()])
    image = transform(image)

    image = image.to(device)
    predicted = final_model([image])
    predicted_score = list(predicted[0]['scores'].detach().
    cpu().numpy())
    predicted_temp = [predicted_score.index(x) for x in
    predicted_score if x>conf][-1]
    masks = (predicted[0]['masks']>0.5).squeeze().detach().
    cpu().numpy()
    # print(pred[0]['labels'].numpy().max())
```

```
predicted_class_val = [CLASSES[i] for i in
list(predicted[0]['labels'].cpu().numpy())]
predicted_box_val = [[(i[0], i[1]), (i[2], i[3])] for i in
list(predicted[0]['boxes'].detach().cpu().numpy())]
masks = masks[:predicted_temp+1]
predicted_class_name = predicted_class_
val[:predicted_temp+1]
predicted_box_score = predicted_box_val[:predicted_temp+1]

return masks, predicted_box_score, predicted_class_name

def segment_image(image_path, confidence=0.5, rect_thickness=2,
text_size=2, text_thickness=2):

    masks_conf, box_conf, predicted_class = generate_
    prediction(image_path, confidence)
    image = cv2.imread(image_path)
    image = cv2.cvtColor(image, cv2.COLOR_BGR2RGB)
    for i in range(len(masks_conf)):
      rgb_mask = get_mask_color(masks_conf[i])
      image = cv2.addWeighted(image, 1, rgb_mask, 0.5, 0)
      cv2.rectangle(image, box_conf[i][0], box_conf[i][1],
      color=(0, 255, 0), thickness=rect_thickness)
      cv2.putText(image,predicted_class[i], box_conf[i]
      [0], cv2.FONT_HERSHEY_SIMPLEX, text_size,
      (0,255,0),thickness=text_thickness)
    plt.figure(figsize=(20,30))
    plt.imshow(image)
    plt.xticks([])
    plt.yticks([])
    plt.show()

segment_image('/content/pedestrian_img.jpg', confidence=0.7)
```

Figure 4-3. *Output on the custom data*

Summary

The chapter discussed how image segmentation works and its variations in the market. This is one important aspect of solving the problem that involves image segmentation. This also established the concept of fine-tuning through an example. Moving forward, the concepts learned here are going to be useful in understanding the concepts of computer vision.

The next chapter looks at how we can build pipelines to help with business problems involving image similarity.

CHAPTER 5

Image-Based Search and Recommendation System

In order to retain and acquire new customers, especially in the e-commerce arena, customer service needs to be top-notch. There are already thousands of e-commerce platforms and the number will only increase in the future. Platforms with excellent customer experience will survive in long term.

The question is how do we provide great customer service? There are many ways we can enhance the customer experience. Making search engines state-of-the-art will not only make customers happy but will also increase sales via cross-selling.

There are many ways to use search engines and recommendation engines with natural language processing, deep learning, etc. The newest edition is image processing. We can use the power of image processing, deep learning, and pretrained models to create image-based search and recommendation systems that produce great results.

© Akshay Kulkarni, Adarsha Shivananda, and Nitin Ranjan Sharma 2022
A. Kulkarni et al., *Computer Vision Projects with PyTorch*,
https://doi.org/10.1007/978-1-4842-8273-1_5

Problem Statement

When users are searching e-commerce platforms, they typically search for product names and descriptions. Suppose you are looking for a blue t-shirt. You can use a basic search phrase to get relevant results. However, say you like a t-shirt that you saw someone wearing at a party and it is blue with white stripes and has a black floral print with a red collar. That's a difficult phrase to search for, and the results will be uninspired. This is where image-based search and recommendation systems come in.

Instead of providing instant recommendations using the user's product descriptions, we can give recommendations based on an image. This will capture more details of the product than a text description, especially in the fashion category.

Approach and Methodology

To search images, we first need to understand how machines analyze these images. We need to convert these images into numbers/vectors. Once we do that, it's a matter of time to solve this problem.

Images can be represented in the form of vectors or embedded data using pretrained models. We can the ResNet18 model from PyTorch, which will convert images into embedded information or vectors for this project.

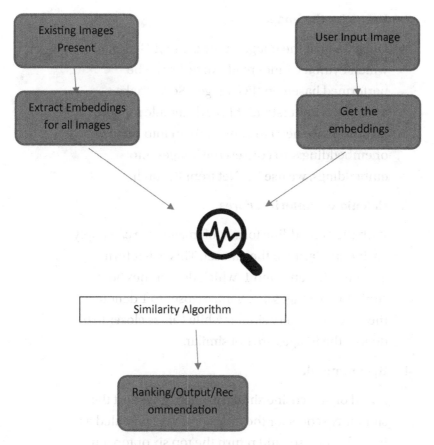

Figure 5-1. *Image similarity pipeline flowchart*

Our aim here is to find similar images based on the input. To achieve this, we need to do the following things, as shown in Figure 5-1:

1. **Import existing images and search image.**

   The first step is to load the images into the working environment. We will use the OpenCV functionalities to do so.

2. **Vectorize the images.**

Once we load the images, they are in JPG format, which Python cannot read. No tasks can be performed based on JPG images. So, to make the images understandable and suitable for our algorithm, we need to convert them into vectors or embeddings. To convert the images into embeddings, we use ResNet from PyTorch.

3. **Calculate similarity scores.**

With the embedding format of images, we can apply cosine similarity to the dataset. This will return a score between 0 and 1, which determines how similar two images are. Values closer to 1 denote the images are very similar while values closer to 0 denote the images are not similar.

4. **Recommend.**

Based on the cosine similarity matrix, we sort the similarity scores for the index or image provided as input by the user and return the top six or top ten similar items.

Implementation

Now that we understand what the problem is and how we can solve it, let's jump into the implementation.

The Dataset

We will be using a famous Kaggle dataset for this use case. Here's the link to download the dataset: https://www.kaggle.com/paramaggarwal/fashion-product-images-small

The structure of the dataset is as follows:

- A folder named images, which contains all the images of items available in the dataset. The image names are in the format [id].jpg where [id] is the item ID provided in CSV.

- A CSV file called styles.csv has ten columns.

The ten columns from styles.csv are defined as follows:

- id: A unique number provided to that particular item

- gender: Gender (unnecessary bias, should avoid it)

- masterCategory: Main category under which the item falls

- subCategory: Sub-category under which the item falls

- articleType: Type of product

- baseColor: Base color of the product

- season: Season it's suitable for

- year: The year that the item was uploaded

- usage: When can the item be used

- productDisplayName: Display name on the web page

Installing and Importing Libraries

We will be using OpenCV and PyTorch vision for this problem. Let's install them.

```
!pip install swifter
!pip install torchvision
!pip install opencv-python

# Importing matplotlib for plotting
import matplotlib.pyplot as plt

# Importing numpy for numerical operations
import numpy as np

# Importing pandas for preprocessing
import pandas as pd

# Importing joblib to dump and load embeddings df
import joblib

# Importing cv2 to read images
import cv2

# Importing cosine_similarity to find similarity between images
from sklearn.metrics.pairwise import cosine_similarity

# Importing flatten from pandas to flatten 2-D array
from pandas.core.common import flatten

# Importing the below libraries for our model building

#import torch
import torch
import torch.nn as nn
```

```
#import cv models
import torchvision.models as models
import torchvision.transforms as transforms
from torch.autograd import Variable

#import image
from PIL import Image

import warnings
warnings.filterwarnings("ignore")
```

Importing and Understanding the Data

Now we will import the data and try to make sense of it. We need to import two things.

✓   Metadata of the images

✓   The images themselves

```
# Importing the metadata
df = pd.read_csv('../fashion-product-images-small/styles.
csv',error_bad_lines=False,warn_bad_lines=False)

#top 10 rows
df.head(10)
```

	id	gender	masterCategory	subCategory	articleType	baseColour	season	year	usage	productDisplayName
0	15970	Men	Apparel	Topwear	Shirts	Navy Blue	Fall	2011.0	Casual	Turtle Check Men Navy Blue Shirt
1	39386	Men	Apparel	Bottomwear	Jeans	Blue	Summer	2012.0	Casual	Peter England Men Party Blue Jeans
2	59263	Women	Accessories	Watches	Watches	Silver	Winter	2016.0	Casual	Titan Women Silver Watch
3	21379	Men	Apparel	Bottomwear	Track Pants	Black	Fall	2011.0	Casual	Manchester United Men Solid Black Track Pants
4	53759	Men	Apparel	Topwear	Tshirts	Grey	Summer	2012.0	Casual	Puma Men Grey T-shirt
5	1855	Men	Apparel	Topwear	Tshirts	Grey	Summer	2011.0	Casual	Inkfruit Mens Chain Reaction T-shirt
6	30805	Men	Apparel	Topwear	Shirts	Green	Summer	2012.0	Ethnic	Fabindia Men Striped Green Shirt
7	26960	Women	Apparel	Topwear	Shirts	Purple	Summer	2012.0	Casual	Jealous 21 Women Purple Shirt
8	29114	Men	Accessories	Socks	Socks	Navy Blue	Summer	2012.0	Casual	Puma Men Pack of 3 Socks
9	30039	Men	Accessories	Watches	Watches	Black	Winter	2016.0	Casual	Skagen Men Black Watch

Figure 5-2. *Input data snapshot*

As we can see in Figure 5-2, the first column is the image's ID. It's the metadata where the image ID along with any information about the image is saved.

Let's look at the different articleTypes and their frequency.

```
#set style
plt.style.use('ggplot')
```

```
# Understanding the data, how many different articleType are
present and knowing their frequency
```

```
plt.figure(figsize=(7,28))
df.articleType.value_counts().sort_values().plot(kind='barh')
```

Figure 5-3 shows the output. We have the highest numbers in the tshirts and shirts categories.

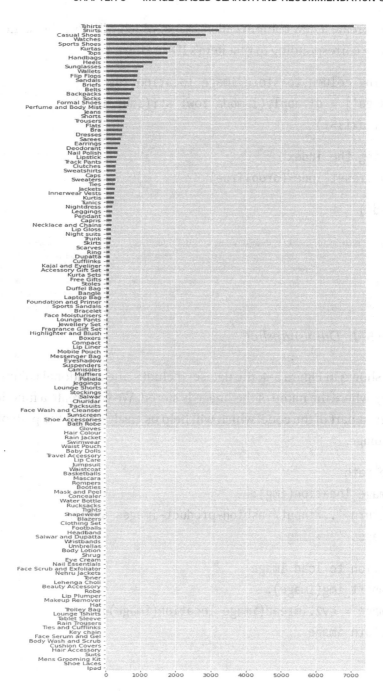

Figure 5-3. *Class distribution*

Let's create a new column called image, which will store the name of the image corresponding to that item's ID.

```
#creating column to store image location ids
df['image'] = df.apply(lambda row: str(row['id']) +
".jpgs, axis=1)
```

```
#reseting the index
df = df.reset_index(drop=True)
```

```
df.head()
```

	Id	gender	masterCategory	subCategory	articleType	baseColour	season	year	usage	productDisplayName	image
0	15970	Men	Apparel	Topwear	Shirts	Navy Blue	Fall	2011.0	Casual	Turtle Check Men Navy Blue Shirt	15970.jpg
1	39386	Men	Apparel	Bottomwear	Jeans	Blue	Summer	2012.0	Casual	Peter England Men Party Blue Jeans	39386.jpg
2	59263	Women	Accessories	Watches	Watches	Silver	Winter	2016.0	Casual	Titan Women Silver Watch	59263.jpg
3	21379	Men	Apparel	Bottomwear	Track Pants	Black	Fall	2011.0	Casual	Manchester United Men Solid Black Track Pants	21379.jpg
4	53759	Men	Apparel	Topwear	Tshirts	Grey	Summer	2012.0	Casual	Puma Men Grey T-shirt	53759.jpg

Figure 5-4. *Data snapshot*

As shown in Figure 5-4, the dataset includes one additional column (image), where the name of the image is stored. We will create a function in the next part of the code, which will help us easily obtain the path of every image.

```
#image path
def image_location(img):
    return '../input/fashion-product-images-small/
    images/'  + img
```

```
# function to load image
def import_img(image):
    image = cv2.imread(image_location(image))
    return image
```

These functions will help us load the image using cv2. Let's also create one more function that can be used to display the images given their row and column names.

```
def show_images(images, rows = 1, cols=1,figsize=(12, 12)):

    #define fig
    fig, axes = plt.subplots(ncols=cols,
    nrows=rows,figsize=figsize)

    #loop for images
    for index,name in enumerate(images):
        axes.ravel()[index].imshow(cv2.cvtColor(images[name],
        cv2.COLOR_BGR2RGB))
        axes.ravel()[index].set_title(name)
        axes.ravel()[index].set_axis_off()

    #plot
    plt.tight_layout()

# Generation of a dictionary of {index, image}
figures = {'im'+str(i): import_img(row.image) for i, row in
df.sample(6).iterrows()}

# Plotting the images in a figure, with 2 rows and 3 columns
show_images(figures, 2, 3)
```

Figure 5-5 shows the images with row = 2 and column = 3.

Figure 5-5. Image output

Feature Engineering

Plotting images is for a better understanding, but as we have discussed
multiple times, we need to convert the images into numbers using
pixels. The images need to be converted into embeddings. We can either
train our embeddings or use the pretrained image models to get better
performance. In this case, we use the ResNet18 PyTorch model to convert
images into feature vectors.

Let's first spend some time understanding what ResNet does.

ResNet18

ResNet18 is a convolutional neural network (CNN). As the number
suggests, it has 18 layers. It's being trained on millions of images that are

extracted from the ImageNet dataset. It has the capacity to classify more than 1000 object types. Figure 5-6 shows the original architecture from ResNet18 paper.

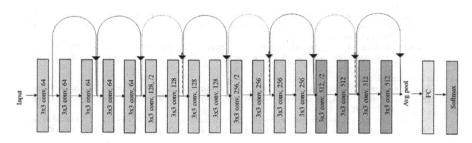

Figure 5-6. *Original architecture of ResNet18*

Let's start with an implementation.

```
# Defining the input shape
width= 224
height= 224

# Loading the pretrained model
resnetmodel = models.resnet18(pretrained=True)

# selecting the layer
layer = resnetmodel._modules.get('avgpool')

#evaluation
resnetmodel.eval()
```

```
ResNet(
  (conv1): Conv2d(3, 64, kernel_size=(7, 7), stride=(2, 2), padding=(3, 3), bias=False)
  (bn1): BatchNorm2d(64, eps=1e-05, momentum=0.1, affine=True, track_running_stats=True)
  (relu): ReLU(inplace=True)
  (maxpool): MaxPool2d(kernel_size=3, stride=2, padding=1, dilation=1, ceil_mode=False)
  (layer1): Sequential(
    (0): BasicBlock(
      (conv1): Conv2d(64, 64, kernel_size=(3, 3), stride=(1, 1), padding=(1, 1), bias=False)
      (bn1): BatchNorm2d(64, eps=1e-05, momentum=0.1, affine=True, track_running_stats=True)
      (relu): ReLU(inplace=True)
      (conv2): Conv2d(64, 64, kernel_size=(3, 3), stride=(1, 1), padding=(1, 1), bias=False)
      (bn2): BatchNorm2d(64, eps=1e-05, momentum=0.1, affine=True, track_running_stats=True)
    )
    (1): BasicBlock(
      (conv1): Conv2d(64, 64, kernel_size=(3, 3), stride=(1, 1), padding=(1, 1), bias=False)
      (bn1): BatchNorm2d(64, eps=1e-05, momentum=0.1, affine=True, track_running_stats=True)
      (relu): ReLU(inplace=True)
      (conv2): Conv2d(64, 64, kernel_size=(3, 3), stride=(1, 1), padding=(1, 1), bias=False)
      (bn2): BatchNorm2d(64, eps=1e-05, momentum=0.1, affine=True, track_running_stats=True)
    )
  )
  (layer2): Sequential(
    (0): BasicBlock(
      (conv1): Conv2d(64, 128, kernel_size=(3, 3), stride=(2, 2), padding=(1, 1), bias=False)
      (bn1): BatchNorm2d(128, eps=1e-05, momentum=0.1, affine=True, track_running_stats=True)
      (relu): ReLU(inplace=True)
      (conv2): Conv2d(128, 128, kernel_size=(3, 3), stride=(1, 1), padding=(1, 1), bias=False)
```

Figure 5-7. *Convolution block*

Figure 5-7 shows the architecture of the model. Now let's extract the embedded vectors for the images and save them in an object.

```
# scaling the data
s_data = transforms.Scale((224, 224))

# normalizing
standardize = transforms.standardize(mean=[0.7, 0.6, 0.3],
                                      std=[0.2, 0.3, 0.1])

# converting to tensor
convert_tensor = transforms.ToTensor()

# creating the missing image object
missing_img = []

#function to get embeddings

def vector_extraction(resnetmodel, image_id):
```

```python
# exception handling to ignore missing images
try:

    img = Image.open(image_location(image_id)).
    convert('RGB')

    t_img = Variable(standardize(convert_tensor(s_
    data(img))).unsqueeze(0))

    embeddings = torch.zeros(512)

    def select_d(m, i, o):
        embeddings.copy_(o.data.reshape(o.data.size(1)))

    hlayer = layer.register_forward_hlayer(select_d)

    resnetmodel(t_img)

    hlayer.remove()
    emb = embeddings

    return embeddings
# If file not found
except FileNotFoundError:
    # Store the index of such entries in missing_img list
    and drop them later
    missed_img = df[df['image']==image_id].index
    print(missed_img)
    missing_img.append(missed_img)
```

This function will load the image, reshape it into 224*224, and convert it to an array, which will later go through the ResNet model. This will return an array of 512 values, which are the 512 feature vectors for that particular image.

Let's apply this function to a sample image and see the output.

```
# Testing if our vector_extraction function works well on
sample image

sample_embedding_0 = vector_extraction(resnetmodel,
df.iloc[0].image)

# Plotting the sample image and its embeddings

img_array = import_img(df.iloc[0].image)

plt.imshow(cv2.cvtColor(img_array, cv2.COLOR_BGR2RGB))
print(img_array.shape)
print(sample_embedding_0)
```

```
(80, 60, 3)
tensor([1.6732e-02, 9.8327e-01, 4.0268e-02, 1.1314e-01, 2.0513e-01, 1.2468e+00,
        3.5904e-02, 3.3680e-01, 1.3279e+00, 4.8053e-01, 4.5403e-02, 2.1866e-01,
        1.2002e+00, 1.2201e-01, 0.0000e+00, 9.9961e-03, 5.6686e-01, 0.0000e+00,
        1.9427e-02, 2.7316e-01, 2.9556e-01, 1.0254e+00, 1.1648e+00, 5.4014e-01,
        2.9776e-02, 1.2624e-01, 5.3572e-01, 2.1451e+00, 1.5348e-01, 3.6843e-01,
        1.1278e+00, 2.5455e-01, 2.3566e-01, 9.0818e-01, 1.4324e+00, 1.0864e+00,
        7.2151e-01, 2.8588e-01, 5.6683e-01, 7.9897e-02, 6.0556e-01, 6.3392e-02,
        2.2239e-01, 1.5460e+00, 2.6952e+00, 0.0000e+00, 4.6124e-02, 2.3475e-02,
        1.3130e+00, 5.5342e-01, 2.3303e+00, 3.7319e-01, 7.1914e-01, 4.4571e-01,
        8.5868e-01, 5.1455e-01, 4.8082e-01, 2.3485e+00, 4.6088e-01, 1.9201e+00,
        3.0348e-01, 7.3000e-01, 8.2374e-01, 5.0691e-01, 1.0031e-01, 3.2392e-02,
        5.1186e-01, 2.9504e-01, 1.7705e-01, 1.4258e+00, 4.5813e-01, 1.8374e+00,
        1.4661e-01, 6.7185e-02, 2.7939e+00, 3.2873e-01, 1.1578e+00, 2.1376e+00,
        4.7114e-01, 4.1420e-01, 1.0309e+00, 1.8506e+00, 3.0370e-02, 2.0246e+00,
        2.5223e+00, 1.1975e-01, 8.8195e-01, 1.7082e-01, 4.0317e+00, 2.5442e+00,
        5.8607e-01, 5.5378e-01, 1.5619e+00, 2.5786e+00, 1.9007e+00, 1.1317e+00,
        6.3828e-01, 1.2285e+00, 4.2008e-01, 1.8927e-01, 5.8589e-02, 2.8445e-01,
        1.5736e-01, 0.0000e+00, 7.2721e-01, 2.5659e+00, 5.7278e+00, 1.6366e-01,
        8.1007e-01, 2.1702e-01, 4.4027e+00, 1.9851e+00, 8.2007e-03, 1.4142e+00,
```

Figure 5-8. *Output tensor from image*

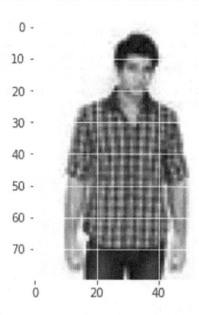

Figure 5-9. *Sample image*

Figures 5-8 and 5-9 show the sample image and its vector for a random sample from the dataset.

```
# Testing if our vector_extraction function works well on
sample image
sample_embedding_1 = vector_extraction(resnetmodel,
df.iloc[1000].image)
```

```
# Plotting the sample image and its embeddings
img_array = import_img(df.iloc[1000].image)
plt.imshow(cv2.cvtColor(img_array, cv2.COLOR_BGR2RGB))
print(img_array.shape)
print(sample_embedding_1)
```

```
1.7549e-01, 2.2049e-01, 1.2595e-01, 5.1752e-01, 1.0725e-01, 5.7015e-01,
1.6123e+00, 8.5854e-02, 2.3698e-01, 5.3520e-01, 7.1433e-01, 6.8964e-01,
2.2787e-01, 2.2046e+00, 1.6270e-01, 4.7040e-01, 3.4608e+00, 8.8849e-01,
3.3635e-01, 5.2675e-01, 2.0664e-01, 1.8067e-01, 2.1441e-02, 7.7205e-01,
2.2574e-01, 2.7452e-01, 0.0000e+00, 4.3661e-02, 4.2577e-01, 2.4761e-01,
1.6707e-01, 1.0226e-01, 4.7133e-02, 7.4051e-01, 1.7953e-01, 3.6949e-01,
1.7816e-01, 8.2362e-01, 0.0000e+00, 1.3964e-01, 2.2376e+00, 2.8166e-01,
6.7271e-04, 0.0000e+00])
```

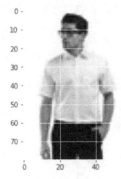

Figure 5-10. *Output tensor*

Figure 5-10 shows another example image and its vector for a random sample from the dataset. When we print emb0 and emb1, we get arrays with 512 values. Those are the feature vectors. We can see that the images correspond to those vectors as well.

Now, let's take these embeddings and find the distance between them using cosine similarity. This will help us determine how similar the items are based on the value.

```
#Finding the similarity between those two images

cos_sim = cosine_similarity(sample_embedding_0.unsqueeze(0),
            sample_embedding_1.unsqueeze(0))

print('\nCosine similarity: {0}\n'.format(cos_sim))

# output
Cosine similarity: [[0.8811257]]
```

The similarity between these two images is 0.88, which means that the images are nearly the same. As we observe the images, both shirts are the same size. That's the reason the similarity is high.

We have extracted the embeddings for only two images. Now let's write a loop and extract the vectors for all the images we have in our dataset.

```
%%time
import swifter

# Applying embeddings on subset of this huge dataset
df_embeddings    = df[:5000] #We can apply on entire df, like:
df_embeddings = df

#looping through images to get embeddings
map_embeddings = df_embeddings['image'].swifter.apply(lambda
img: vector_extraction(resnetmodel, img))

#convert to series
df_embs        = map_embeddings.apply(pd.Series)
print(df_embs.shape)
df_embs.head()
```

```
Pandas Apply:    0%|        | 0/5000 [00:00<?, ?it/s]
(5000, 512)
CPU times: user 10min 49s, sys: 3.65 s, total: 10min 53s
Wall time: 5min 44s
```

	0	1	2	3	4	5	6	7	8	9	...	502	503	504	505	506
0	0.016732	0.983267	0.040268	0.113140	0.205126	1.246753	0.035904	0.336803	1.327888	0.480529	...	0.584026	0.483292	1.229778	0.738920	0.000000
1	0.034120	0.804465	0.071094	0.286108	0.118644	0.485673	0.767113	0.116924	1.131223	1.229429	...	0.125503	0.554490	0.160279	0.211642	0.000000
2	0.306779	0.196791	2.325818	0.337869	0.206403	0.410262	2.865744	0.493546	2.894568	3.824196	...	0.377007	3.216576	2.293661	1.343940	1.047547
3	0.052566	0.312828	0.318465	0.045758	0.207992	0.486139	0.871359	0.437957	0.861973	1.257671	...	0.000401	0.126220	0.117900	0.174461	0.000000
4	0.146032	0.624985	0.023857	0.201500	0.273301	2.073840	0.038832	0.537267	1.338017	0.428539	...	0.039452	1.069758	0.774631	0.874319	0.000000

5 rows × 512 columns

Figure 5-11. Embedding snapshot

We get feature vectors for the first 5000 images. It took a long time to get these feature vectors. To save time, we will save this embedding in our local system, which can be imported during future use.

There are different ways to save:

- Use the df.to_csv( ) function from pandas
- Use the joblib.dump( ) function from joblib

```
#export the embeddings
df_embs.to_csv('df_embs.csv')

# importing the embeddings
df_embs = pd.read_csv('df_embs.csv')
df_embs.drop(['Unnamed: 0','index'],axis=1,inplace=True)
df_embs.dropna(inplace=True)

#exporting as pkl
joblib.dump(df_embs, 'df_embs.pkl', 9)

#importing the pkl
df_embs = joblib.load('df_embs.pkl')
```

Calculating Similarity and Ranking

Now that we have vectors for each image, we can find the similarity scores between them and then rank them to get recommendations.

Cosine similarity between two vectors is calculated as shown in Figure 5-12.

$$similarity(A,B) = \frac{A \cdot B}{\|A\| \times \|B\|} = \frac{\sum_{i=1}^{n} A_i \times B_i}{\sqrt{\sum_{i=1}^{n} A_i^2} \times \sqrt{\sum_{i=1}^{n} B_i^2}}$$

Figure 5-12. *Formula to calculate cosine similarity*

```
# Calculating similarity between images ( using embedding values )
cosine_sim = cosine_similarity(df_embs)
```

```
# Previewing first 4 rows and 4 columns similarity just to
check the structure of cosine_sim
cosine_sim[:4, :4]
```

```
#output
array([[1.0000007 , 0.76683545, 0.5455518 , 0.779508  ],
       [0.76683545, 1.0000002 , 0.49617064, 0.88492715],
       [0.5455518 , 0.49617064, 0.9999991 , 0.52310663],
       [0.779508  , 0.88492715, 0.52310663, 1.000001  ]],
       dtype=float32)
```

Now that we have the similarity matrix, let's define a function that gives recommendations based on the cosine similarity score. We need to keep the following three things as inputs to the function to get the desired recommendations:

- Image ID

- Metadata dataset name

- Number of recommendations required

```
# Storing the index values in a series index_vales for
recommending
```

```
index_vales = pd.Series(range(len(df)), index=df.index)
index_vales
```

```
# Defining a function that gives recommendations based on the
cosine similarity score
```

```
def recommend_images(ImId, df, top_n = 6):
```

```
# Assigning index of reference into sim_ImId
sim_ImId    = index_vales[ImId]

# Storing cosine similarity of all other items with item
requested by user in sml_scr as a list
sml_scr = list(enumerate(cosine_sim[sim_ImId]))

# Sorting the list of sml_scr
sml_scr = sorted(sml_scr, key=lambda x: x[1], reverse=True)

# Extracting the top n values from sml_scr
sml_scr = sml_scr[1:top_n+1]

# ImId_rec will return the index of similar items
ImId_rec    = [i[0] for i in sml_scr]

# ImId_sim will return the value of similarity score
ImId_sim    = [i[1] for i in sml_scr]

return index_vales.iloc[ImId_rec].index, ImId_sim
```

We have created a function that takes the index of the image for which we want similar images, the dataframe, and an integer that determines the number of images to be recommended as the parameters.

When we pass these three parameters into this function, it returns the index of the top n similar images and their similarity scores.

```
# Sample given below
recommend_images(3810, df, top_n = 5)
```

```
#output
(Int64Index([2400, 3899, 3678, 4818, 2354], dtype='int64'),
 [0.9632292, 0.9571406, 0.95574236, 0.9539639, 0.95376974])
```

Returning only the index value and similarity scores is not enough; therefore, we need to visualize the recommendations as well, by plotting the images in recommended indices.

Visualizing the Recommendations

We have arrived at the most interesting part of the whole chapter, the results. Let's create a function to visualize these recommendations and then evaluate them.

```
def Rec_viz_image(input_imageid):

    # Getting recommendations
    idx_rec, idx_sim = recommend_images(input_imageid, df,
    top_n = 6)

    # Printing the similarity score
    print (idx_sim)

    # Plotting the image of item requested by user
    plt.imshow(cv2.cvtColor(import_img(df.iloc[input_imageid].
    image), cv2.COLOR_BGR2RGB))

    # Generating a dictionary of { index, image }
    figures = {'im'+str(i): import_img(row.image) for i, row in
    df.loc[idx_rec].iterrows()}

    # Plotting the similar images in a figure, with 2 rows and
    3 columns
    show_images(figures, 2, 3)
```

The visualize function will call the recommend_images function and store the returned indices and scores. With the indices, stored images will be plotted using the plot_figures function.

Let's look at some examples. The first image is the image at index 3810, and what follows are the top six similar items.

```
Rec_viz_image(3810)
[0.9632292, 0.9571406, 0.95574236, 0.9539639, 0.95376974,
0.9536929]
```

Figure 5-13 shows the input image.

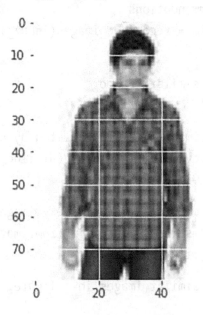

Figure 5-13. *Input image*

Figure 5-14 shows the recommendations for the input image. As we can see, the input image is a shirt and the recommendations we received are also shirts. They all have the same pattern but are different colors. These results are exceptional and realistic.

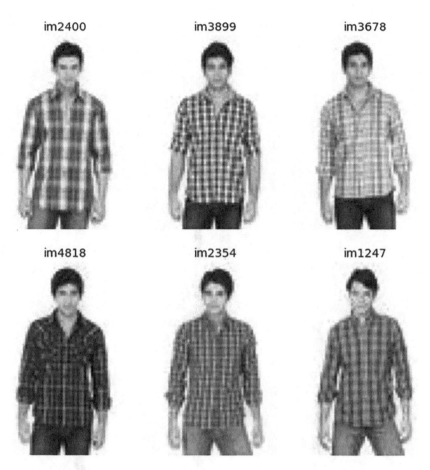

Figure 5-14. *Output images*

For the next sample, we chose a tie. Figure 5-15 shows the output of our function, which shows six images of the tie with similarity scores over 90%.

```
Rec_viz_image(2518)
[0.9506557, 0.931319, 0.928721, 0.9247011, 0.9215533, 0.917436]
```

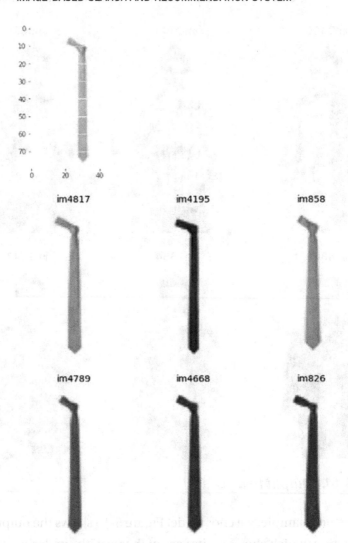

Figure 5-15. *Sample output images*

So far, we have tested on images that are already present in the dataset. Now we will take input images from the user and try to find items similar to their image.

Taking Image Input from Users and Recommending Similar Products

In this function, we will load the image from the path provided by the user and use the ResNet50 model to get its feature vectors. Next, we will find the cosine similarity between the user's image and the rest of the feature vectors in our dataset.

Later, we will sort the cosine similarity scores and select the highest ten. We will extract the similarity score and the indices of the top ten images. The similarity score will be printed so the user can see the items, while the indices are used to plot the images. Here, we are recommending the top ten items, so the figures are listed two rows by five columns.

Let's build the function.

```
def recm_user_input(image_id):

    # Loading image and reshaping it
    img = Image.open('../input/testset-for-image-similarity/' +
    image_id).convert('RGB')

    t_img = Variable(standardize(convert_tensor(s_data(img))).
    unsqueeze(0))

    embeddings = torch.zeros(512)

    def select_d(m, i, o):
        embeddings.copy_(o.data.reshape(o.data.size(1)))

    hlayer = layer.register_forward_hlayer(select_d)

    resnetmodel(t_img)

    hlayer.remove()
    emb = embeddings
```

```python
# Calculating Cosine Similarity
cs = cosine_similarity(emb.unsqueeze(0),df_embs)
cs_list = list(flatten(cs))
cs_df = pd.DataFrame(cs_list,columns=['Score'])
cs_df = cs_df.sort_values(by=['Score'],ascending=False)

# Printing Cosine Similarity
print(cs_df['Score'][:10])

# Extracting the index of top 10 similar items/images
top10 = cs_df[:10].index
top10 = list(flatten(top10))
images_list = []
for i in top10:
    image_id = df[df.index==i]['image']
    images_list.append(image_id)
images_list = list(flatten(images_list))

# Plotting the image of item requested by user
img_print = Image.open('../input/testset-for-image-
similarity/' + image_id)
plt.imshow(img_print)

# Generating a dictionary { index, image }
figures = {'im'+str(i): Image.open('../input/fashion-
product-images-small/images/' + i) for i in images_list}
```

```
# Plotting the similar images in a figure, with 2 rows and
  3 columns
fig, axes = plt.subplots(2, 5, figsize = (8,8) )
for index,name in enumerate(figures):
    axes.ravel()[index].imshow(figures[name])
    axes.ravel()[index].set_title(name)
    axes.ravel()[index].set_axis_off()
plt.tight_layout()
```

Let's download an image from Google and use it as input.

```
recm_user_input('test5.jpg')
4036    0.824246
954     0.810449
3268    0.808926
4528    0.808186
3299    0.807687
295     0.806027
1978    0.805003
2900    0.803676
3688    0.800311
1229    0.800130
Name: Score, dtype: float64
```

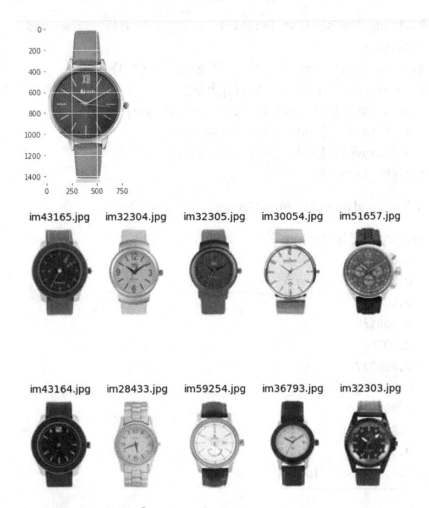

Figure 5-16. *Sample output images*

At the top of Figure 5-16 is a sample image of a watch that has been downloaded from Google and is not a part of the dataset. Next, we passed the watch as the input image, and the recommendations are all watches. Further, we can classify with our human senses that six-seven watches are more feminine like the sample watch. Also, the similarity scores are nearly 70%. So the output looks good.

Let's try another image, this time of a shoe. Figure 5-17 shows the results.

```
Recm_user_input('test14.jpg')
```

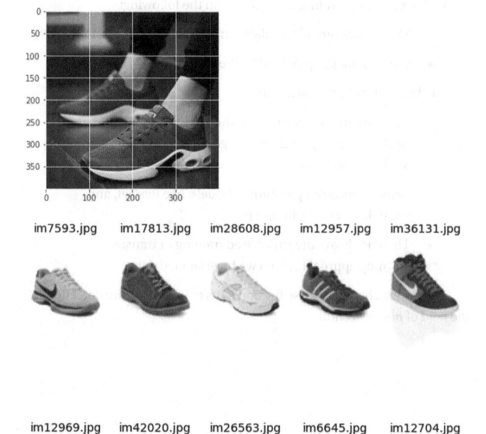

Figure 5-17. *Recommendations from the model for shoes*

Summary

We applied a pretrained ResNet18 model to our dataset to recommend and search images. The results look pretty good.

We tried recommending items based on the following:

- The images already available in our dataset

- Custom images provided by the user

Further scope of improvements:

- Recommend items based on the `articleType`, `masterCategory`, and `subCategory` features of the `styles.csv` dataset.

- Try various other pretrained models, like ResNet, and see if the accuracy improves.

- There is always the supervised training or transfer learning approach, given we have labeled data.

In the next chapter, we use the concepts of image feature detection in the field of *pose detection*.

CHAPTER 6

Pose Estimation

Human pose estimation (HPE) is a computer vision task that detects human poses by estimating major keypoints, such as eyes, ears, hands, and legs, in a given frame/video. Figure 6-1 shows an example of human pose estimation in action.

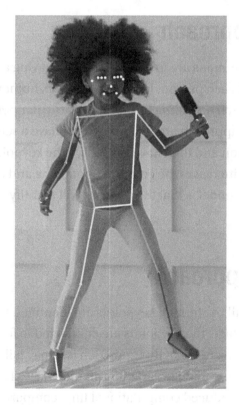

Figure 6-1. *HPE example*

© Akshay Kulkarni, Adarsha Shivananda, and Nitin Ranjan Sharma 2022
A. Kulkarni et al., *Computer Vision Projects with PyTorch*,
https://doi.org/10.1007/978-1-4842-8273-1_6

Human body pose detection helps track human body parts and joints. Some of the keypoints to identify in a human body are the arms, legs, eyes, ears, nose, etc., which can help us track movements.

HPE is mostly widely used in robotics, understanding human activity and behavior, sports analytics, and so on.

Deep learning concepts, especially the CNN architecture, is tailored and designed particularly for HPE.

There are two approaches to this problem:

- Top-down approach

- Bottom-up approach

Top-Down Approach

Using this approach, humans are first identified by drawing an estimated boundary box around each person. In the second step, human keypoints are identified inside each bounding box for that particular person. Disadvantages of this approach are that we need to have a separate model for human identification and then have to identify the keypoints inside all bounding boxes. This increases the computational time and complexity. The advantage of this model is that the network will identify all the humans in the frame.

Bottom-Up Approach

Using this approach, all the human keypoints are identified first in a given frame. In the second stage, the keypoints are connected to form a human-like skeleton. The disadvantage of this approach is it may fail to recognize smaller scale humans due to scale variations in the image. The advantage of this approach is the reduced computational time compared to the top-down approach.

Here are the more common HPE models used today:

- OpenPose: 2019

- HRNet: 2019

- Higher HRNet: 2020

- AlphaPose: 2018

- Mask R-CNN: 2018

- Dense pose: 2018

- DeepCut: 2016

- DeepPose: 2014

- Pose Net: 2015

OpenPose

OpenPose is a real-time, multi-person, multi-stage pose estimation algorithm built on VGG19 as its backbone. This algorithm follows a bottom-up approach. The input image is sent to the VGG-19 network for extracting the feature maps. The extracted features maps are passed to multistage CNN. Each stage contains two branches that run parallelly.

Branch-1

This branch creates heat maps/confidence maps for the keypoints to detect. A separate heat map is generated for all the keypoints.

Branch-2

This branch creates part affinity fields (PAFs). PAFs can identify connections between the keypoints.

The output from the two branches is mapped to identify the right connections using line integral. L2 loss is calculated between the predicted result (heat map, PAF) and the ground truth (heat map, PAF). Two L2 loss functions are used—one at the end of each branch. During training, the overall loss is calculated as the sum of these two loss functions.

Output from Stage 1 is passed to Stage 2 to improvise the results. The depth of the model increases by increasing the number of stages. Since there can be multiple persons in the images, weighted Bipartite graphs are used to connect the parts of the same person. The connected pairs are merged to form a human skeleton. The model can detect up to 135 keypoints on a single image. Figures 6-2 and 6-3 show the architecture and flow chart of OpenPose.

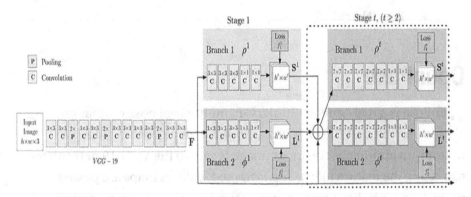

Figure 6-2. *Architecture of OpenPose*

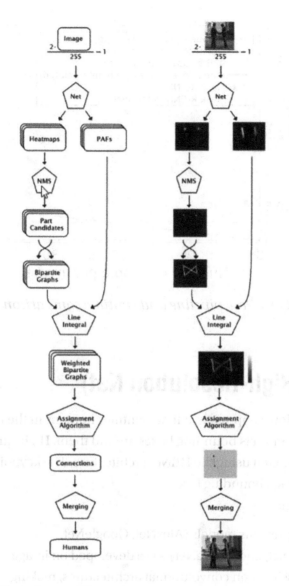

Figure 6-3. *Flowchart of OpenPose*

Figure 6-4 shows the OpenPose runtime in comparison to other models. Default OpenPose and max accuracy OpenPose promise better performance.

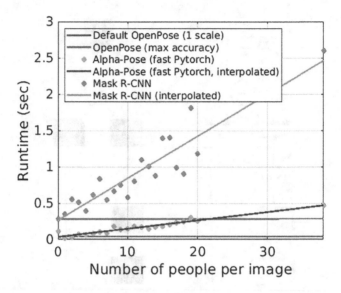

Figure 6-4. *OpenPose pipeline and runtime comparison to other models*

HRNet (High-Resolution Net)

This is a top-down approach. It first identifies humans in the image using Faster-RCNN and sets bounding boxes around them. High-quality feature maps are generated using the HRNet architecture. The keypoints are then identified in each bounding box.

Motivation:

1. All previous models (AlexNet, GoogleNet, ResNet, and DenseNet) were developed on image classification convolutional architectures, making the output low resolution and position-insensitive. The low resolution can be increased in these architectures using dilated convolutions, but computational time increases.

2. Up-sampling is an alternative to this problem. The
    U-net, SegNet, DeConvnet, and Hourglass models
    use the up-sampling technique. In this technique,
    Stage 1 input images are converted to low resolution
    for classification. In Stage 2, the high-resolution
    images will be recovered from low-resolution by
    sequentially connected convolutions. But the
    complete recovery of HR from LR is not possible and
    position sensitivity of representation is weak.

HRNet is a universal architecture for visual recognition. Its architecture
is not based on any classification networks, which use convolutions
in series. In the HRNet architecture, multiple resolution convolutions
are connected in parallel with repeated fusions using up-sampling and
down-sampling techniques. This network maintains HR representations
from start to end. Repeated fusions among the resolutions strengthen
the high- and low-resolution representations. HR convolutions are
converted to LR convolutions using a down-sampling technique called
"stride convolution." LR convolutions are converted to HR convolutions
using the "bilinear up-sampling" technique. The HR branch preserves
spatial information and the LR branch preserves contextual information.
Figures 6-5, 6-6, and 6-7 show HRNet's detailed architecture.

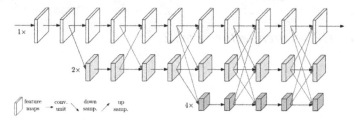

High-resolution networks (HRNet): Connect multi-resolution convolutions in *parallel*
with *repeated fusions*

Figure 6-5. *HRNet architecture, part 1*

Major observations:

- In classification, the convolutions are placed in series. But in HRNet, convolutions are placed parallel.

- For up-sampling, a bilinear function is used instead of convolution (due to time complexity).

- Strided convolution is used to down-sample the HR images (to avoid information loss).

The number of blocks in stages 2, 3, and 4 are 1 ,4, and 3. These numbers are not well optimized (as per the author). Parameter and computational complexity are not higher than ResNet due to the reduced number of channels in HRNet. Since the architecture is a multi-resolution network, the output is delivered in all resolutions (high, medium, and low). For HPE, only the HR channel output is used. For semantic segmentation and face alignment, all resolution outputs are used.

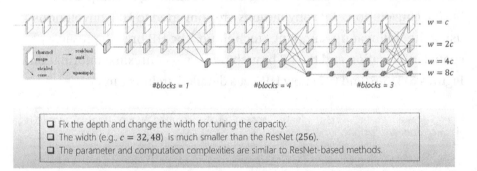

Figure 6-6. *HRNet architecture, part 2*

Relation to Regular Convolution

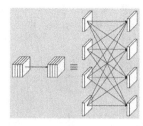

Regular convolution

Multi-resolution convolution
(across-resolution fusion)

Figure 6-7. *HRNet architecture, part 3*

Higher HRNet

This is a bottom-up approach, unlike the original HRNet model. The major issue with previous bottom-up approaches is dealing with scale variations (such as children or people far way). This issue is solved in the Higher HRNet model by using HR feature maps (from HRNet) and HR heat maps (using a deconvolution step).

This network is built using the HRNet architecture as the backbone. The input image is passed to a stem (containing two convolution blocks, which reduces the resolution to ¼). Later the image is passed through the HRNet architecture to produce the HR feature maps. The HR feature maps are fed to the deconvolution blocks. These deconvolution blocks (feature maps and predicted heat maps from HRNet) take as input and will produce two HR heat maps, followed by four residual blocks (batch norm + ReLU) to up-sample the feature maps. This model uses a high-resolution supervision technique to train the model. The ground truth keypoints are transformed to all resolution heat maps to generate the ground truth heat maps. The predicted heat maps are validated against the ground truth heat maps, to calculate the loss (mean squared error). Figure 6-8 shows the architecture diagram.

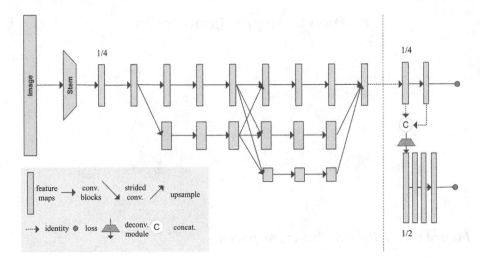

Figure 6-8. *Higher HRNet architecture*

From theoretical studies, Higher HRNet shows promising results in addressing computational time (using the bottom-up approach) and the scale variation problem (using multi-resolution).

PoseNet

PoseNet is a pose estimator built on tensorflow.js that runs on mobile devices. It estimates the pose of the human body by detecting human body points such as eyes, nose, mouth, wrists, elbows, hips, knees, etc. It forms a skeleton-like structure of the pose by joining these keypoints.

It works for both single and multiple human pose detection.

How Does PoseNet Work?

PoseNet was trained with ResNet and MobileNet models. The ResNet model has higher accuracy. But it is large and has many layers, which makes it slower. Thus, it is good to go with the MobileNet model, as it is designed to run on mobile devices. Pose estimation happens in two phases:

- An input RGB image is fed into a convolutional neural network.

- A single-pose or multi-pose algorithm is used to obtain keypoints (coordinates) and their confidence scores from the model outputs.

The output of PoseNet model is a pose object that contains a list of keypoints and confidence scores for each detected person. Figure 6-9 shows pose vs keypoint confidence.

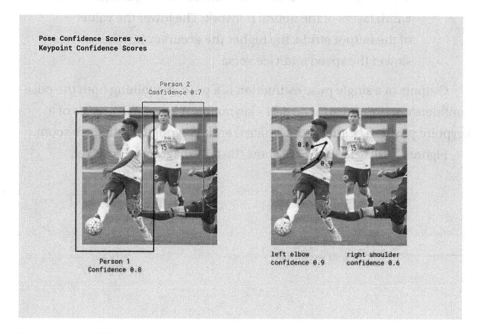

Figure 6-9. *Illustration of pose vs keypoint confidence*

Single Person Pose Estimation

This is the case when only one person is centered in an input image or video. Inputs for the single pose estimation algorithm are as follows:

- **Input image element:** An input image element that the program will predict the pose for.

- **Image scale factor:** A number between 0.2 and 1. By default, it is set to 0.5.

- **Flip horizontal:** By default, this is set to false. If the poses have to be flipped horizontally/vertically, then this has to be set to true. When a video is by default flipped horizontally, the poses are returned to the proper orientation.

- **Output stride:** This should be 32, 16, or 8. By default, it is set to 16. This variable influences the height and width layers of the neural network. The lower the value of the output stride, the higher the accuracy but the slower the speed and vice versa.

Outputs of a single pose estimation is a pose, containing both the pose confidence score and an array of 17 keypoints. A keypoint consists of a keypoint position (x and y coordinates) and a keypoint confidence score.

Figures 6-10, 6-11, and 6-12 show the flow diagram of PoseNet.

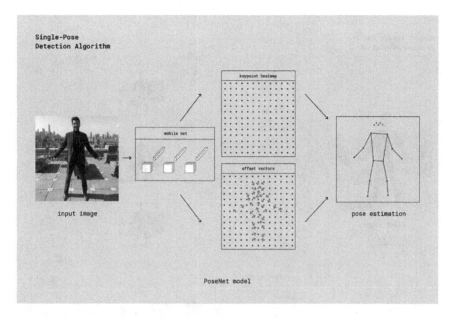

Figure 6-10. *The flow diagram of PoseNet*

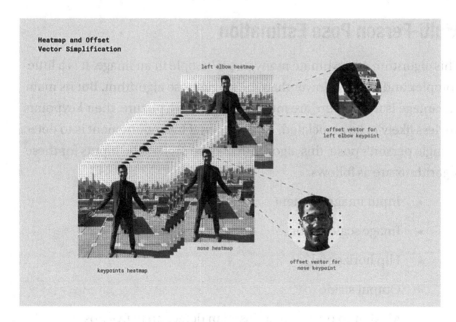

Figure 6-11. *The flow diagram of PoseNet, part 2*

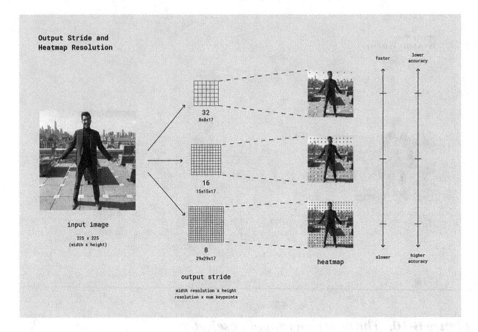

Figure 6-12. *The flow diagram of PoseNet, part 3*

Multi-Person Pose Estimation

This algorithm can estimate many poses/people in an image. It is a little complex and slightly slower than the single-pose algorithm. But its main advantage is that if there are multiple people in a picture, their keypoints are less likely to be associated. Hence, even if the requirement is to detect a single person's pose, this algorithm may be preferable. Inputs for these algorithms are as follows:

- Input image element

- Image scale factor

- Flip horizontal

- Output stride

- Maximum pose detections: Can detect up to five poses

- Pose confidence threshold

- Non-maximum suppression (NMS) radius: This controls the minimum distance between poses that are returned. Its default value is 20.

The output of this algorithm is an array of poses. Each pose contains 17 keypoints along with each keypoint's score.

Pros and Cons of PoseNet

Consider these pros and cons of PoseNet:

- As it is a lightweight model, it can be used for mobile/edge devices.

- The single pose estimation algorithm associates the keypoints with the wrong person if there is more than one human in the picture.

Applications of Pose Estimation

The following are common applications of pose estimation:

- Human activity recognition

- Human fall detection

- Motion tracking for consoles

- Training robots

Test Cases Performed Retail Store Videos

Case 1: Tested PoseNet model with 1080p resolution video of one hour with an fps of 2. Outcome:

- CPU Utilization: 80-90%

- Memory: 1.2 to 1.5GB

- FPS: 15

- Processing time and DB insertion for a one-hour video: 20 to 25 minutes

Case 2: Tested PoseNet model with 720p and 480p resolution videos of one hour with an fps of 2. Outcome:

- For 720p, processing time and DB insertion for one-hour video: 8 to 10 minutes with 16 fps

- For 480p, processing time and DB insertion for one-hour video: 4 to 5 minutes with 25 fps

Implementation

Now that we have covered some of the theoretical aspects and models, let's move on to the implementation part, using one of the methods and the pretrained model. What follows is a step-by-step guide to detecting a human body pose for a single image using PyTorch.

We are going to use the "Keypoint-RCNN Using ResNet-50 Architecture with Feature pyramid Network" solution for human pose and keypoint detection. The code is separated into seven blocks for easy understanding. Here are the steps:

1. Identify the list of human keypoints to track.

2. Identify the possible connections between the keypoints.

3. Load the pretrained model from the PyTorch library.

4. Input image preprocessing and modeling.

5. Build custom functions to plot the output (keypoints and skeleton).

6. Plot the output on the input image.

First, let's import the required libraries:

```
#Import libraries
import os
import numpy as np

#For importing keypoint RCNN pretrained model and image
preprocessing
import torchvision
import torch

#For reading the image
import cv2

#For visualization
import matplotlib.pyplot as plt

#Mount google drive
#Change directory to the respective folder containing
images folder
from google.colab import drive
drive.mount('/content/drive')
%cd '/content/drive/MyDrive/Colab Notebooks/Bodypose'
```

Step 1: Identify the List of Human Keypoints to Track

A list of human keypoints can be found in Figure 6-13. These keypoints are the target entities in the deep learning model, which are discussed in Step 3.

3D KEYPOINTS AND THEIR SPECIFICATION

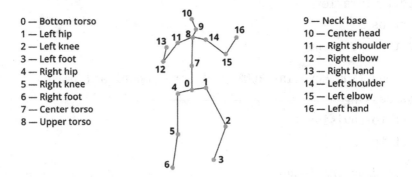

0 — Bottom torso
1 — Left hip
2 — Left knee
3 — Left foot
4 — Right hip
5 — Right knee
6 — Right foot
7 — Center torso
8 — Upper torso

9 — Neck base
10 — Center head
11 — Right shoulder
12 — Right elbow
13 — Right hand
14 — Left shoulder
15 — Left elbow
16 — Left hand

Figure 6-13. *Illustration of human keypoints*

Figure 6-13 shows the illustration of human keypoints.

```
# List of keypoints for humans (count=17)
human_keypoints = ['nose','left_eye','right_eye','left_
ear','right_ear','left_shoulder','right_shoulder','left_elbow',
             'right_elbow','left_wrist','right_wrist','left_
hip','right_hip','left_knee', 'right_knee', 'left_
ankle','right_ankle']

print(human_keypoints)

#output
['nose', 'left_eye', 'right_eye', 'left_ear', 'right_ear',
 'left_shoulder', 'right_shoulder', 'left_elbow', 'right_elbow',
```

```
'left_wrist', 'right_wrist', 'left_hip', 'right_hip', 'left_
knee', 'right_knee', 'left_ankle', 'right_ankle']
```

Step 2: Identify the Possible Connections Between the Keypoints

Now identify the possible connections between the keypoints. For example, the left ear has a connection to the left eye. All the possible connections can be found in the following code snippet.

```python
# Possible connections between the human key points to form a
  structure
def possible_keypoint_connections(keypoints):
    connections = [
        [keypoints.index('right_eye'), keypoints.
        index('nose')],
        [keypoints.index('right_eye'), keypoints.
        index('right_ear')],
        [keypoints.index('left_eye'), keypoints.index('nose')],
        [keypoints.index('left_eye'), keypoints.
        index('left_ear')],
        [keypoints.index('right_shoulder'), keypoints.
        index('right_elbow')],
        [keypoints.index('right_elbow'), keypoints.
        index('right_wrist')],
        [keypoints.index('left_shoulder'), keypoints.
        index('left_elbow')],
        [keypoints.index('left_elbow'), keypoints.index('left_
        wrist')],
        [keypoints.index('right_hip'), keypoints.index('right_
        knee')],
```

```
    [keypoints.index('right_knee'), keypoints.index('right_
    ankle')],
    [keypoints.index('left_hip'), keypoints.index('left_
    knee')],
    [keypoints.index('left_knee'), keypoints.index('left_
    ankle')],
    [keypoints.index('right_shoulder'), keypoints.
    index('left_shoulder')],
    [keypoints.index('right_hip'), keypoints.
    index('left_hip')],
    [keypoints.index('right_shoulder'), keypoints.
    index('right_hip')],
    [keypoints.index('left_shoulder'), keypoints.
    index('left_hip')]
    ]
  return connections

connections = possible_keypoint_connections(human_keypoints)
```

Step 3: Load the Pretrained Model from the PyTorch Library

In this blog, we are using the PyTorch pretrained model keypoint-RCNN with ResNet50 architecture for keypoint detection. Load the model with this parameter: (pretrained= True).

```
# create a model from the pretrained keypointrcnn_resnet50_
fpn class
pretrained_model = torchvision.models.detection.keypointrcnn_
resnet50_fpn(pretrained=True)
```

```
# call the eval() method to prepare the model for
inference mode.
```

```
pretrained_model.eval()
#output
Downloading: "https://download.pytorch.org/models/keypointrcnn_
resnet50_fpn_coco-fc266e95.pth" to /root/.cache/torch/hub/
checkpoints/keypointrcnn_resnet50_fpn_coco-fc266e95.pth
100%
226M/226M [00:04<00:00, 15.1MB/s]
KeypointRCNN(
  (transform): GeneralizedRCNNTransform(
      Normalize(mean=[0.485, 0.456, 0.406], std=[0.229,
      0.224, 0.225])
      Resize(min_size=(640, 672, 704, 736, 768, 800), max_
      size=1333, mode='bilinear')
  )
```

Step 4: Input Image Preprocessing and Modeling

The original image needs to be normalized before passing to the model. Normalization is performed using the transforms.Compose() and transforms.ToTensor() classes from the transforms module of TorchVision. Place the input image into the images folder in the current working directory.

```
# import the transforms module
from torchvision import transforms as T

# Read the image using opencv
img_path = "images/image1.JPG"
img = cv2.imread(img_path)

# preprocess the input image
transform = T.Compose([T.ToTensor()])
img_tensor = transform(img)
```

```
# forward-pass the model
output = pretrained_model([img_tensor])[0]

print(output.keys())
#output
dict_keys(['boxes', 'labels', 'scores', 'keypoints',
'keypoints_scores'])
```

Figure 6-14 is the image we used as input.

Figure 6-14. *Input image*

Step 5: Build Custom Functions to Plot the Output

Build custom functions to plot the predicted keypoints and the skeleton of the body (by connecting the keypoints).

```
#Functions to plot keypoints and skeleton of input image
```

```python
def plot_keypoints(img, all_keypoints, all_scores, confs,
keypoint_threshold=2, conf_threshold=0.9):
    # initialize a set of colors from the rainbow spectrum
    cmap = plt.get_cmap('rainbow')
    # create a copy of the image
    img_copy = img.copy()
    # pick a set of N color-ids from the spectrum
    color_id = np.arange(1,255, 255//len(all_keypoints)).
    tolist()[::-1]
    # iterate for every person detected
    for person_id in range(len(all_keypoints)):
      # check the confidence score of the detected person
      if confs[person_id]>conf_threshold:
        # grab the keypoint-locations for the detected person
        keypoints = all_keypoints[person_id, ...]
        # grab the keypoint-scores for the keypoints
        scores = all_scores[person_id, ...]
        # iterate for every keypoint-score
        for kp in range(len(scores)):
            # check the confidence score of detected keypoint
            if scores[kp]>keypoint_threshold:
                # convert the keypoint float-array to a python-
                list of intergers
                keypoint = tuple(map(int, keypoints[kp, :2].
                detach().numpy().tolist()))
                # pick the color at the specific color-id
                color = tuple(np.asarray(cmap(color_id[person_
                id]))[:-1])*255)
                # draw a cirle over the keypoint location
                cv2.circle(img_copy, keypoint, 30, color, -1)

    return img_copy
```

```python
def plot_skeleton(img, all_keypoints, all_scores, confs,
keypoint_threshold=2, conf_threshold=0.9):

    # initialize a set of colors from the rainbow spectrum
    cmap = plt.get_cmap('rainbow')
    # create a copy of the image
    img_copy = img.copy()
    # check if the keypoints are detected
    if len(output["keypoints"])>0:
      # pick a set of N color-ids from the spectrum
      colors = np.arange(1,255, 255//len(all_keypoints)).
      tolist()[::-1]
      # iterate for every person detected
      for person_id in range(len(all_keypoints)):
          # check the confidence score of the detected person
          if confs[person_id]>conf_threshold:
            # grab the keypoint-locations for the
            detected person
            keypoints = all_keypoints[person_id, ...]

            # iterate for every limb
            for conn_id in range(len(connections)):
              # pick the start-point of the limb
              limb_loc1 = keypoints[connections[conn_id][0], :2]
              .detach().numpy().astype(np.int32)
              # pick the start-point of the limb
              limb_loc2 = keypoints[connections[conn_id][1], :2]
              .detach().numpy().astype(np.int32)
              # consider limb-confidence score as the minimum
              keypoint score among the two keypoint scores
```

```
        limb_score = min(all_scores[person_id,
        connections[conn_id][0]], all_scores[person_id,
        connections[conn_id][1]])
        # check if limb-score is greater than threshold
        if limb_score> keypoint_threshold:
            # pick the color at a specific color-id
            color = tuple(np.asarray(cmap(colors[person_
            id]))[:-1])*255)
            # draw the line for the limb
            cv2.line(img_copy, tuple(limb_loc1),
            tuple(limb_loc2), color, 25)

    return img_copy
```

Step 6: Plot the Output on the Input Image

Using the custom functions from Step 5, plot the predicted keypoints and
skeleton onto the original image.

```
#Key points
keypoints_img = plot_keypoints(img, output["keypoints"],
output["keypoints_scores"], output["scores"],keypoint_
threshold=2)

cv2.imwrite("output/keypoints-img.jpg", keypoints_img)

plt.figure(figsize=(8, 8))
plt.imshow(keypoints_img[:, :, ::-1])
plt.show()

#Output
```

Figure 6-15 shows the image with keypoints.

Figure 6-15. *Image with keypoints*

```
#Skeleton
skeleton_img = plot_skeleton(img, output["keypoints"],
output["keypoints_scores"], output["scores"],keypoint_
threshold=2)

cv2.imwrite("output/skeleton-img.jpg", skeleton_img)

plt.figure(figsize=(8, 8))
plt.imshow(skeleton_img[:, :, ::-1])
plt.show()

#plot
```

Figure 6-16 shows the image as a skeleton.

Figure 6-16. *Image as a skeleton*

Here are the outcomes from other images we tried for more than one person. You can find these on this book's Git link as well.

Figure 6-17. *Image with keypoints*

Figure 6-18. *Image as a skeleton*

Summary

This chapter explored the architecture and code walkthrough for developing a pose estimator model. This is widely used in the industry.

Are you now confident that you can build an application that acts as a "virtual gym instructor"? It's worth giving it a try, given how many people want a home gym. In the next chapter, we explore how to do anomaly detection on images.

CHAPTER 7

Image Anomaly Detection

The study of machine learning has put us in the course of studying various patterns and behavior. It has allowed us to build models that can study closed environments. Predictive power often follows the model training process. It is an important question that we need to ask often when we are training a model. There is another question that needs an answer—how much data is sufficient to help the model understand the distribution such that we can have a good representation? This chapter will work out an example and the concepts regarding these important questions. We are discussing anomaly detection in computer vision.

We have a machine learning model that learns the data distribution and eventually can be used to make predictions about the unknown dataset. The process of learning is restricted to the distribution represented by the data we are using for training. After the training process is finished, a few samples might contradict the majority behavior. We have to note that detecting anomalies is subject to perspectives, such as how lenient the distribution needs to be. For example, a polished steel sheet can have rows of straight lines from the machines. There can be light scratches and these still might not be considered defective (anomalies). In other scenarios, these scratches or abruptness might be considered anomalies. So, for all scenarios, anomalies need to have a threshold.

© Akshay Kulkarni, Adarsha Shivananda, and Nitin Ranjan Sharma 2022
A. Kulkarni et al., *Computer Vision Projects with PyTorch*,
https://doi.org/10.1007/978-1-4842-8273-1_7

Anomaly detection has a lot of applications in images, for example, detecting anomalies in a metal sheet at a construction site. Anomaly detection can be used to find abnormalities on a conveyor belt.

Anomaly Detection

Anomaly detection in visual analytics, like in all other domains, can be divided into two major types:

- **Novelty detection:** During the training process, the models are subjected to data that has resulted from a standard event distribution. When we are testing or predicting for unknown samples, the algorithm is supposed to find anomalous data. In this process, it is assumed the data does not have any non-standard data. This is an example of a semi-supervised learning method.

- **Outlier detection:** The algorithm, in this case, is subjected to both standard and non-standard data. Since by principle the standard data will be concentrated, the algorithm learns it and disregards the outliers. We can take an example of a decision tree, wherein the branch will always try to separate outliers earlier in the split process. The data in this approach is polluted with standard and non-standard data. The algorithm figures out which data points are inliers and which are outliers. This is an unsupervised way of training.

We have multiple ways to detect outliers or novelty. We can use statistical methods such as population mean and standard deviation to find outliers. However, in those scenarios, the knowledge of distribution is a must. In machine learning methods, we have a few algorithms that can help us with anomaly detection.

- **Local outlier factor:** The algorithm calculates a value that quantifies the local density deviation. It tries to locate samples that have lower density when compared to its neighbors.

- **Isolation forest:** There are ways to use iterative splitting based on decisions, which can be used to determine outliers in samples. If we can leverage the algorithm based on the ensemble of decision trees or random forest, we can easily conclude that the samples that fall on shorter paths of the random forest are the anomalies.

- **One class SVM:** It can be thought to be an extension of the SVM class, wherein upon deciding on a threshold, the support of a probability distribution can be checked, thus separating the outliers in the process.

Let's look at some of the applications in the field of computer vision.

- **Unsupervised density estimation:** The algorithm tries to estimate the probability distribution of the features or of the training images. Once the distribution is known by the model, for all unknown samples it tries to determine the difference of the sample from the distribution.

- **Unsupervised image reconstruction:** A general process of training an encoder-decoder architecture. It gets the network to learn the vectorized latent features and reconstruct the original image, with some loss. The reconstruction loss for normal images will be smaller compared to the anomalies.

- **One class anomaly detection:** This approach is similar to the one-class SVM discussed earlier. The algorithm tries to estimate a decision boundary to separate the normal class from the anomalies.

Generative classes of algorithms can be used to detect anomalies and have been established by multiple researchers. Now that we have worked out some of the basic concepts, let's look at an example of anomaly detection.

Some of the approaches include the following:

- Use a pretrained model with training on the last few layers for anomaly classification

- The encoding and decoding approach

- Anomaly image classification and locating the anomaly in an image using feature maps

Approach 1: Using a Pretrained Classification Model

Finding an anomaly image from a given image dataset can be considered as a binary image classification task, i.e., whether the image is an anomaly or not based on the training dataset. One of the proven architectures called VGG-16 is used here to train the last few layers.

The VGG-16 architecture contains 16 layers, 13 of which are convolution layers and the last three are fully connected layers. This network is trained to predict the class of the input out of a total of 1,000 classes.

In the present approach, pretrained weights are used for the first ten convolutional layers. The last layers are used to model-train the custom dataset. The output is classified into one of two classes. The highlighted box in Figure 7-1 is used to train a custom dataset.

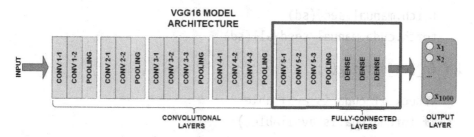

Figure 7-1. *The VGG-16 architecture*

Step 1: Import the Required Libraries

```
#import torch
import torch
import torchvision
import matplotlib.pyplot as plt

#import time,os etc
import time
import os
import numpy as np
import random
from distutils.version import LooseVersion as Version
from itertools import product
```

Step 2: Create the Seed and Deterministic Functions

These functions help generate the same random number for all iterations.

```
def seed_setting(sd):
    os.environ["PL_GLOBAL_SEED"] = str(sd)
    random.seed(sd)
    np.random.seed(sd)
    torch.manual_seed(sd)
    torch.cuda.manual_seed_all(sd)

def fn_det_setting():

    #check if cuda is available
    if torch.cuda.is_available():
        torch.backends.cudnn.benchmark = False
        torch.backends.cudnn.deterministic = True

    #check torch version
    if torch.__version__ <= Version("1.7"):
        torch.fn_det_setting(True)
    else:
        torch.use_deterministic_algorithms(True)
```

Step 3: Set the Hyperparameter

```
#set seed, batch size
RNDM_SEED = 245
btch_input_sz = 128
epch_nmbr = 25
DEVICE = torch.device('cuda:1' if torch.cuda.is_available()
else 'cpu')
```

```
seed_setting(RNDM_SEED)
```

```
#fn_det_setting() This may not work on Gpu because some
algorithms are not deterministic on Gpu./
```

Step 4: Import the Dataset

Here is the training data:

```
tr_ds_path = "/content/drive/MyDrive/car_img/tr"
#Training Images
```

Here is the validation data:

```
vd_ds_path = "/content/drive/MyDrive/car_img/vds"
#Validation Images
```

Here is the test data:

```
ts_ds_path = "/content/drive/MyDrive/car_img/ts" #Test Images
```

Step 5: Image Preprocessing Stage

Image transformation includes these steps:

- Image resizing: To maintain common image sizes in the train, test, and validation datasets

- Image cropping: To crop the images' edges

- Converting images to tensors: For PyTorch implementation

- Image normalization: For faster loss convergence

```
import torch.utils.data as data

tr_data_trans = torchvision.transforms.Compose([
    torchvision.transforms.Resize((70, 70)),
    torchvision.transforms.RandomCrop((64, 64)),
    torchvision.transforms.ToTensor(), #it converts data in the
    range 0-255 to 0-1.
    torchvision.transforms.Normalize((0.485, 0.456, 0.406),
    (0.229, 0.224, 0.225))])

validation_data_trans = torchvision.transforms.Compose([
    torchvision.transforms.Resize((70, 70)),
    torchvision.transforms.CenterCrop((64, 64)),
    torchvision.transforms.ToTensor(),
    torchvision.transforms.Normalize((0.485, 0.456, 0.406),
    (0.229, 0.224, 0.225))])

tst_data_transform = torchvision.transforms.Compose([
    torchvision.transforms.Resize((70, 70)),
    torchvision.transforms.CenterCrop((64, 64)),
    torchvision.transforms.ToTensor(),
    torchvision.transforms.Normalize((0.485, 0.456, 0.406),
    (0.229, 0.224, 0.225))])
```

DataLoader functions help by passing the data in parallel, thus making the data-loading process faster.

```
train_ds_cln = torchvision.datasets.ImageFolder(root=tr_ds_
path, transform= tr_data_trans)
train_loader_cln = data.DataLoader(train_ds_cln, btch_input_
sz=206, shuffle=True)
```

```
test_ds_cln = torchvision.datasets.ImageFolder(root=ts_ds_path,
transform= tst_data_transform)
test_loader_cln = data.DataLoader(test_ds_cln, btch_input_
sz=206, shuffle=True)

valid_ds_cln = torchvision.datasets.ImageFolder(root=vd_ds_
path, transform= validation_data_trans)
valid_loader_cln = data.DataLoader(valid_ds_cln, btch_input_
sz=63, shuffle=True)

# Checking the dataset
for images, labels in train_loader_cln:
    print('Image batch dimensions:', images.shape)
    print('Image label dimensions:', labels.shape)
    print('Class labels of 10 examples:', labels[:10])
    break
```

Here is the output:

```
Image batch dimensions: torch.Size([206, 3, 64, 64])
Image label dimensions: torch.Size([206])
Class labels of 10 examples: tensor([1, 0, 0, 1, 0, 0, 1,
1, 1, 1])
```

Here is the training dataset:

```
for images, labels in train_loader_cln:
    print('Image batch dimensions:', images.shape)
    print('Image label dimensions:', labels.shape)
    print('Class labels of 10 examples:', labels[:10])
    break
```

Here is the output:

```
Image batch dimensions: torch.Size([206, 3, 64, 64])
Image label dimensions: torch.Size([206])
Class labels of 10 examples: tensor([0, 1, 0, 1, 1, 1, 1,
1, 0, 1])
```

```
tr_ds = images
tr_ds.shape
```

Here is the output:

```
torch.Size([206, 3, 64, 64])
```

```
tr_label = labels
tr_label.shape
```

Here is the output:

```
torch.Size([206])
```

Here is the validation dataset:

```
for images, labels in valid_loader_cln:
    print('Image batch dimensions:', images.shape)
    print('Image label dimensions:', labels.shape)
    print('Class labels of 10 examples:', labels[:10])
    break
```

Here is the output:

```
Image batch dimensions: torch.Size([63, 3, 64, 64])
Image label dimensions: torch.Size([63])
Class labels of 10 examples: tensor([1, 1, 0, 0, 1, 0, 1,
1, 1, 1])
```

```
vd_ds = images
vd_ds.shape
```

Here is the output:

```
torch.Size([63, 3, 64, 64])
```

Here is the test dataset:

```
for images, labels in test_loader_cln:
    print('Image batch dimensions:', images.shape)
    print('Image label dimensions:', labels.shape)
    print('Class labels of 10 examples:', labels[:10])
    break
```

Here is the output:

```
Image batch dimensions: torch.Size([63, 3, 64, 64])
Image label dimensions: torch.Size([63])
Class labels of 10 examples: tensor([1, 1, 1, 1, 1, 0, 1,
1, 1, 0])
```

Step 6: Load the Pretrained Model

```
model = torchvision.models.vgg16(pretrained=True)
model
```

Here is the output:

```
VGG(
  (features): Sequential(
    (0): Conv2d(3, 64, kernel_size=(3, 3), stride=(1, 1),
        padding=(1, 1))
    (1): ReLU(inplace=True)
    (2): Conv2d(64, 64, kernel_size=(3, 3), stride=(1, 1),
        padding=(1, 1))
    (3): ReLU(inplace=True)
```

```
(4): MaxPool2d(kernel_size=2, stride=2, padding=0,
     dilation=1, ceil_mode=False)
(5): Conv2d(64, 128, kernel_size=(3, 3), stride=(1, 1),
     padding=(1, 1))
(6): ReLU(inplace=True)
(7): Conv2d(128, 128, kernel_size=(3, 3), stride=(1, 1),
     padding=(1, 1))
(8): ReLU(inplace=True)
(9): MaxPool2d(kernel_size=2, stride=2, padding=0,
     dilation=1, ceil_mode=False)
(10): Conv2d(128, 256, kernel_size=(3, 3), stride=(1, 1),
      padding=(1, 1))
(11): ReLU(inplace=True)
(12): Conv2d(256, 256, kernel_size=(3, 3), stride=(1, 1),
      padding=(1, 1))
(13): ReLU(inplace=True)
(14): Conv2d(256, 256, kernel_size=(3, 3), stride=(1, 1),
      padding=(1, 1))
(15): ReLU(inplace=True)
(16): MaxPool2d(kernel_size=2, stride=2, padding=0,
      dilation=1, ceil_mode=False)
(17): Conv2d(256, 512, kernel_size=(3, 3), stride=(1, 1),
      padding=(1, 1))
(18): ReLU(inplace=True)
(19): Conv2d(512, 512, kernel_size=(3, 3), stride=(1, 1),
      padding=(1, 1))
(20): ReLU(inplace=True)
(21): Conv2d(512, 512, kernel_size=(3, 3), stride=(1, 1),
      padding=(1, 1))
(22): ReLU(inplace=True)
(23): MaxPool2d(kernel_size=2, stride=2, padding=0,
      dilation=1, ceil_mode=False)
```

```
    (24): Conv2d(512, 512, kernel_size=(3, 3), stride=(1, 1),
          padding=(1, 1))
    (25): ReLU(inplace=True)
    (26): Conv2d(512, 512, kernel_size=(3, 3), stride=(1, 1),
          padding=(1, 1))
    (27): ReLU(inplace=True)
    (28): Conv2d(512, 512, kernel_size=(3, 3), stride=(1, 1),
          padding=(1, 1))
    (29): ReLU(inplace=True)
    (30): MaxPool2d(kernel_size=2, stride=2, padding=0,
          dilation=1, ceil_mode=False)
  )
  (avgpool): AdaptiveAvgPool2d(output_size=(7, 7))
  (classifier): Sequential(
    (0): Linear(in_features=25088, out_features=4096,
         bias=True)
    (1): ReLU(inplace=True)
    (2): Dropout(p=0.5, inplace=False)
    (3): Linear(in_features=4096, out_features=4096, bias=True)
    (4): ReLU(inplace=True)
    (5): Dropout(p=0.5, inplace=False)
    (6): Linear(in_features=4096, out_features=1000, bias=True)
  )
)
```

Step 7: Freeze the Model

Here, the adaptive AVG pooling layer is the bridge between the convolutional layers and linear layers. We are going to train only linear layers. The easiest way is to freeze the whole model first. So we iterate over all the parameters in the model.

Assume we want to fine-tune (train) the last three layers:

```
for param in model.parameters():
    param.requires_grad = False
```

Now we can still run the model forward and backward but it will not update parameters. We fine-tune the last three layers in the next steps.

```
model.classifier[1].requires_grad = True
model.classifier[3].requires_grad = True
```

For the last layer, because the number of class labels differs from ImageNet, we replace the output layer with your output layer:

```
model.classifier[6] = torch.nn.Linear(4096, 2)
```

Step 8: Train the Model

Here's the training process:

```
def find_acc_metric(input_model, input_data_ldr, dvc):
    with torch.no_grad():
        correct_pred, num_examples = 0, 0
        for i, (features, targets) in enumerate(input_
        data_ldr):
            features = features.to(dvc)
            targets = targets.float().to(dvc)

            preds = input_model(features)
            _, predicted_labels = torch.max(preds, 1)

            num_examples += targets.size(0)
            correct_pred += (predicted_labels == targets).sum()
    return correct_pred.float()/num_examples * 100
```

```python
def mdl_training(model, epch_nmbr, train_loader,
                 valid_loader, test_loader, optimizer,
                 device, logging_interval=50,
                 scheduler=None,
                 scheduler_on='valid_acc'):

    tme_strt = time.time()
    list_from_loss, accuracy_train, accuracy_validation =
    [], [], []

    for epoch in range(epch_nmbr):

        model.train()
        for batch_idx, (features, targets) in enumerate(train_
        loader): #iterating over minibatches

            features = features.to(device) #Loading the data
            targets = targets.to(device)

            # ## FORWARD AND BACK PROP
            preds = model(features) #prediction
            loss = torch.nn.functional.cross_entropy(preds,
            targets)
            optimizer.zero_grad()

            loss.backward() #backpropagation

            # ## UPDATE MODEL PARAMETERS
            optimizer.step()

            # ## LOGGING
            list_from_loss.append(loss.item())
            if not batch_idx % logging_interval:
                print(f'Epoch: {epoch+1:03d}/{epch_nmbr:03d} '
```

```
                    f'| Batch {batch_idx:04d}/{len(train_
                        loader):04d} '
                    f'| Loss: {loss:.4f}')

        model.eval()
        with torch.no_grad():  # save memory during inference
            train_acc = find_acc_metric(model, train_loader,
            device=device)
            valid_acc = find_acc_metric(model, valid_loader,
            device=device)
            print(f'Epoch: {epoch+1:03d}/{epch_nmbr:03d} '
                    f'| Train: {train_acc :.2f}% '
                    f'| Validation: {valid_acc :.2f}%')
            accuracy_train.append(train_acc.item())
            accuracy_validation.append(valid_acc.item())

        tr_time = (time.time() - tme_strt)/60
        print(f'Time tr_time: {tr_time:.2f} min')

        if scheduler is not None:

            if scheduler_on == 'valid_acc':
                scheduler.step(accuracy_validation[-1])
            elif scheduler_on == 'minibatch_loss':
                scheduler.step(list_from_loss[-1])
            else:
                raise ValueError(f'Invalid `scheduler_on`
                choice.')

    tr_time = (time.time() - tme_strt)/60
    print(f'Final Training Time: {tr_time:.2f} min')
```

```
    test_acc = find_acc_metric(model, test_loader, device=device)
    print(f'Test accuracy {test_acc :.2f}%')

    return list_from_loss, accuracy_train, accuracy_validation
```

Step 9: Evaluate the Model

```
def Viz_acc(acc_training, val_acc, loc_res):

    epch_nmbr = len(acc_training)

    plt.plot(np.arange(1, epch_nmbr+1),
            acc_training, label='Training')
    plt.plot(np.arange(1, epch_nmbr+1),
            val_acc, label='Validation')

    plt.xlabel('# of Epoch')
    plt.ylabel('Accuracy')
    plt.legend()

    plt.tight_layout()

    if loc_res is not None:
        image_path = os.path.join(
            loc_res, 'plot_acc_training_validation.pdf')
        plt.savefig(image_path)
DEVICE = "cuda" if torch.cuda.is_available() else "cpu"
model = model.to(DEVICE)

optimizer = torch.optim.SGD(model.parameters(), momentum=0.9,
lr=0.01)
scheduler = torch.optim.lr_scheduler.
ReduceLROnPlateau(optimizer,
```

```
                                              factor=0.1,
                                              mode='max',
                                              verbose=True)

list_from_loss, accuracy_train, accuracy_validation = mdl_
training(
    model=model,
    epch_nmbr=5,
    train_loader=train_loader_cln,
    valid_loader=valid_loader_cln,
    test_loader=test_loader_cln,
    optimizer=optimizer,
    device=DEVICE,
    scheduler=scheduler,
    scheduler_on='valid_acc',
    logging_interval=100)
```

Here is the output:

```
Epoch: 001/005 | Batch 0000/0001 | Loss: 1.4587
Epoch: 001/005 | Train: 79.13% | Validation: 76.21%
Time elapsed: 0.34 min
Epoch: 002/005 | Batch 0000/0001 | Loss: 0.8952
Epoch: 002/005 | Train: 92.72% | Validation: 90.29%
Time elapsed: 0.67 min
Epoch: 003/005 | Batch 0000/0001 | Loss: 0.3280
Epoch: 003/005 | Train: 97.57% | Validation: 96.60%
Time elapsed: 0.99 min
Epoch: 004/005 | Batch 0000/0001 | Loss: 0.1774
Epoch: 004/005 | Train: 99.03% | Validation: 96.60%
Time elapsed: 1.32 min
```

```
Epoch: 005/005 | Batch 0000/0001 | Loss: 0.0581
Epoch: 005/005 | Train: 99.51% | Validation: 98.06%
Time elapsed: 1.66 min
Total Training Time: 1.66 min
Test accuracy 100.00%
```

Visualization for training vs validation:

```
Viz_acc(accuracy_train=accuracy_train,
            accuracy_validation=accuracy_validation,
            results_dir=None)
plt.ylim([60, 100])
plt.show()
```

The output is shown in Figure 7-2.

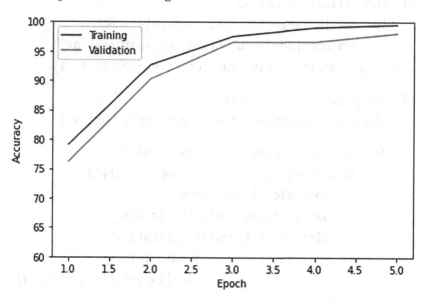

Figure 7-2. *Output training vs validation accuracy*

```
def example_sample(model, data_loader, unnormalizer=None,
class_dict=None):

    for batch_idx, (features, targets) in enumerate(data_loader):

        with torch.no_grad():
            features = features
            targets = targets
            preds = model(features)
            predictions = torch.argmax(preds, dim=1)
        break

    fig, axes = plt.subplots(nrows=3, ncols=5,
                             sharex=True, sharey=True)

    if unnormalizer is not None:
        for idx in range(features.shape[0]):
            features[idx] = unnormalizer(features[idx])
    nhwc_img = np.transpose(features, axes=(0, 2, 3, 1))

    if nhwc_img.shape[-1] == 1:
        nhw_img = np.squeeze(nhwc_img.numpy(), axis=3)

        for idx, ax in enumerate(axes.ravel()):
            ax.imshow(nhw_img[idx], cmap='binary')
            if class_dict is not None:
                ax.title.set_text(f'P: {class_
                dict[predictions[idx].item()]}'
                                  f'\nT: {class_
                                  dict[targets[idx].item()]}')
            else:
                ax.title.set_text(f'P: {predictions[idx]} |
                T: {targets[idx]}')
            ax.axison = False
```

```python
    else:
        for idx, ax in enumerate(axes.ravel()):
            ax.imshow(nhwc_img[idx])
            if class_dict is not None:
                ax.title.set_text(f'P: {class_
                dict[predictions[idx].item()]}'
                                  f'\nT: {class_
                                  dict[targets[idx].item()]}')
            else:
                ax.title.set_text(f'P: {predictions[idx]} |
                T: {targets[idx]}')
            ax.axison = False
    plt.tight_layout()
    plt.show()

class UnNormalize(object): #for plotting the images
    def __init__(self, mean, std):
        self.mean = mean
        self.std = std

    def __call__(self, tensor):
        """

        Parameters:
        ------------

        tensor (Tensor): Tensor image of size (C, H, W) to be
        normalized.

        Returns:
        ------------

        Tensor: Normalized image.
        """

        for t, m, s in zip(tensor, self.mean, self.std):
            t.mul_(s).add_(m)
        return tensor
```

```
model.cpu()
unnormalizer = UnNormalize((0.485, 0.456, 0.406), (0.229,
0.224, 0.2255))
class_dict = {0: 'anomaly',
              1: 'clean'}

example_sample(model=model, data_loader=test_loader_cln,
unnormalizer=unnormalizer, class_dict=class_dict)
```

The output is shown in Figure 7-3.

Figure 7-3. *Output images*

The confusion matrix is as follows:

```
def conf_matrix(model, input_data_ldr, input_dvc):

    trgt_data, pred_data = [], []
    with torch.no_grad():
```

```
    for i, (features, targets) in enumerate(input_
    data_ldr):

        features = features.to(input_dvc)
        targets = targets
        preds = model(features)
        _, predicted_labels = torch.max(preds, 1)
        trgt_data.extend(targets.to('cpu'))
        pred_data.extend(predicted_labels.to('cpu'))
pred_data = pred_data
pred_data = np.array(pred_data)
trgt_data = np.array(trgt_data)

lable_values = np.unique(np.concatenate((trgt_data,
pred_data)))
if lable_values.shape[0] == 1:
    if lable_values[0] != 0:
        lable_values = np.array([0, lable_values[0]])
    else:
        lable_values = np.array([lable_values[0], 1])
n_labels = lable_values.shape[0]
lst = []
z = list(zip(trgt_data, pred_data))
for combi in product(lable_values, repeat=2):
    lst.append(z.count(combi))
mat = np.asarray(lst)[:, None].reshape(n_labels, n_labels)
return mat

def plot_confusion_matrix(conf_mat,
                          hide_spines=False,
                          hide_ticks=False,
                          figsize=None,
```

```
                        cmap=None,
                        colorbar=False,
                        show_absolute=True,
                        show_normed=False,
                        class_names=None):

    if not (show_absolute or show_normed):
        raise AssertionError('Both show_absolute and show_
        normed are False')
    if class_names is not None and len(class_names) !=
    len(conf_mat):
        raise AssertionError('len(class_names) should be equal
        to number of'
                        'classes in the dataset')

    total_samples = conf_mat.sum(axis=1)[:, np.newaxis]
    normed_conf_mat = conf_mat.astype('float') / total_samples

    fig, ax = plt.subplots(figsize=figsize)
    ax.grid(False)
    if cmap is None:
        cmap = plt.cm.Blues

    if figsize is None:
        figsize = (len(conf_mat)*1.25, len(conf_mat)*1.25)

    if show_normed:
        matshow = ax.matshow(normed_conf_mat, cmap=cmap)
    else:
        matshow = ax.matshow(conf_mat, cmap=cmap)

    if colorbar:
        fig.colorbar(matshow)
```

```python
for i in range(conf_mat.shape[0]):
    for j in range(conf_mat.shape[1]):
        cell_text = ""
        if show_absolute:
            cell_text += format(conf_mat[i, j], 'd')
            if show_normed:
                cell_text += "\n" + '('
                cell_text += format(normed_conf_mat[i, j],
                '.2f') + ')'
        else:
            cell_text += format(normed_conf_mat[i, j], '.2f')
        ax.text(x=j,
                y=i,
                s=cell_text,
                va='center',
                ha='center',
                color="white" if normed_conf_mat[i, j] >
                0.5 else "black")

if class_names is not None:
    tick_marks = np.arange(len(class_names))
    plt.xticks(tick_marks, class_names, rotation=90)
    plt.yticks(tick_marks, class_names)

if hide_spines:
    ax.spines['right'].set_visible(False)
    ax.spines['top'].set_visible(False)
    ax.spines['left'].set_visible(False)
    ax.spines['bottom'].set_visible(False)
ax.yaxis.set_ticks_position('left')
ax.xaxis.set_ticks_position('bottom')
```

```
    if hide_ticks:
        ax.axes.get_yaxis().set_ticks([])
        ax.axes.get_xaxis().set_ticks([])

    plt.xlabel('predicted label')
    plt.ylabel('true label')
    return fig, ax

mat = conf_matrix(model=model, data_loader=test_loader_cln,
device=torch.device('cpu'))
plot_confusion_matrix(mat, class_names=class_dict.values())
plt.show()
```

The output is shown in Figure 7-4.

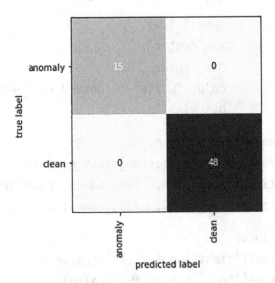

Figure 7-4. *Confusion matrix*

Approach 2: Using Autoencoder

Build an autoencoder training network, which contains the following:

- **Encoder**: To encode the original image (based on pixel values).

- **Decoder**: Reconstruct the image based on the output from the encoder.

Evaluate the model between the original and reconstructed image. Based on the error metric score, the most anomalous digit is detected.

Here is the five-step implementation:

Step 1: Prepare the dataset object.

Step 2: Build the autoencoder network.

Step 3: Train the autoencoder network.

Step 4: Calculate the reconstruction loss based on the original data.

Step 5: Select the most anomalous digit based on the error metric score.

Step 1: Prepare the Dataset Object

Load the input.csv file. The input.csv file contains 65 values for each record. The first 64 values represent grayscale pixel values for the handwritten digits. The last value represents the original class of the digit, which will be between 0 to 9.

Convert the CSV data records to tensors. After that, normalize the pixel values and the original class data.

```
# Step 1
# Preparing Dataset object
  print("\nLoad csv data, convert to  data as normalized
  tensors ")
  #Load .csv dataset containing 65 values
```

```
#first 64 represents 64 pixels values in grayscale format
#last 1 value represent actual digit (in between 0 to 9)
csv_data = "hand_written_digits.txt"

#Convert the csv format to a normalized tensors using helper
function "tensor_converter"
tensor_data = tensor_converter(csv_data)
```

Step 2: Build the Autoencoder Network

Here we build the autoencoder network, which contains the encoder and decoder architecture.

- The encoder converts the original digit pixel values to a lower dimension space. (For example, it converts the 64-pixel grayscale values to 8 values.)

- The decoder reconstructs the original digit from the lower dimension. (For example, it reconstructs the 64-pixel grayscale image using 8 values.)

Fully connected layers are used during encoding and decoding in this problem statement. An encoding network is built using three fully connected (FC) layers, where the first FC layer converts 65 values (64+1) to 48 and the second layer converts 48 values to 32. The last layer converts 32 values to 8. Encoding —> 65–48–32–8.

The decoding network is built using three fully connected layers, where the first layer converts 8 values to 32 and the second FC layer converts 32 values to 48 values. The last layer converts 48 values to 65. Decoding —> 8-32-48-65.

```
def __init__(self):
    super(Autoencoder, self).__init__()
    self.fc1 = T.nn.Linear(65, 48)
    self.fc2 = T.nn.Linear(48, 32)
```

```
    self.fc3 = T.nn.Linear(32, 8)
    self.fc4 = T.nn.Linear(8, 32)
    self.fc5 = T.nn.Linear(32, 48)
    self.fc6 = T.nn.Linear(48, 65)

def encode(self, x):
    # 65-48-32-8
    z = T.tanh(self.fc1(x))
    z = T.tanh(self.fc2(z))
    z = T.tanh(self.fc3(z))
    return z

def decode(self, x):
    # 8-32-48-65
    z = T.tanh(self.fc4(x))
    z = T.tanh(self.fc5(z))
    z = T.sigmoid(self.fc6(z))
    return z
```

Step 3: Train the Autoencoder Network

Train the autoencoder network with hyperparameters such as learning rate, epochs, batch size, loss metrics, and loss optimizer. For training, a helper function takes the autoencoder network, tensor data, and all other hyperparameters, as specified previously.

```
# Step 3. train autoencoder model
    batch_size = 10
    max_epochs = 200
    log_interval = 8
    learning_rate = 0.002

    train(autoenc,tensor_data, batch_size, max_epochs,log_
    interval, learning_rate)
```

Step 4: Calculate the Reconstruction Loss Based on the Original Data

Evaluate the trained model by comparing the original handwritten digit and the reconstructed digit. Calculate and store the image reconstruction loss:

```
#Set autoencoder mode for evaluation
autoenc.eval()

#Store the reconstruction MSE loss
MSE_list = make_err_list(autoenc, tensor_data)
#Sort the list based on MSE loss from highest to lowest
MSE_list.sort(key=lambda x: x[1], reverse=True)
```

Step 5: Select the Most Anomalous Digit Based on the Error Metric Score

Based on the highest MSE loss, we need to find the digit that's an anomaly within the dataset.

```
# Step 5. Show most anomalous handwritten digit in the dataset
    print("Anomaly digit in the dataset given based on
    Highest MSE: ")
    (idx,MSE) = MSE_list[0]
    print(" Index : %4d , MSE : %0.4f" % (idx, MSE))
    display_digit(tensor_data, idx)
```

Output

```
Anomaly digit in the dataset given based on Highest MSE:
Index :  486 , MSE : 0.1360
```

The output is shown in Figure 7-5.

```
digit =   7
```

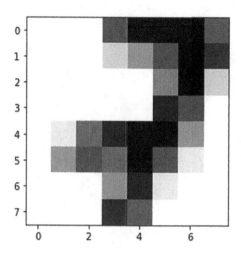

Figure 7-5. *Anomaly output*

Summary

We used a VGG architecture to determine anomalies in sample image datasets. We went through the code and developed an end-to-end pipeline. This model can be used on industry-grade problems with very few changes.

Now that we understand anomaly detections, the next chapter discusses the state-of-the-art use case, which is "image super-resolution." We have seen many apps that improve image quality and resolution. Can you do that yourself by building a model in PyTorch? Let's figure it out in the next chapter.

CHAPTER 8

Image Super-Resolution

With the advent of high-resolution image capturing agents, the information captured in images is huge. Technology has moved from ultra HD to 4K and 8K resolutions. Movies are using high-resolution frames these days; however, there are also situations when they need to enhance a low-resolution image to a high-resolution one. Imagine a scene where the protagonist of a movie is trying to determine the license plate captured from a picture of a speeding car. Super-resolution can now help us zoom into an image to a high degree without distorting it. A few interesting advancements have happened in the industry and we are going to discuss those with some examples.

The existing information in an image cannot be increased from whatever is there initially. In computer science, we have "garbage-in, garbage-out," and this is a similar concept. We cannot expect to find something that's not already there in the image. So, in a way, super-resolution seems far-fetched and quite constrained by information theory. Even so, active research has shown that this problem can be solved.

Let's dive into the problem at hand. We have so far dealt with a supervised form of learning, wherein there is always a loss function associated with the ground truth. The model learns from defined input (X) and expected output (Y). The whole essence of training a model is to help map the inputs to output. But this doesn't happen in unsupervised

© Akshay Kulkarni, Adarsha Shivananda, and Nitin Ranjan Sharma 2022
A. Kulkarni et al., *Computer Vision Projects with PyTorch*,
https://doi.org/10.1007/978-1-4842-8273-1_8

learning. The unsupervised way helps the model learn the pattern in the input data without mapped output. The model learns the patterns in the data and structures its weights around it and then identifies similarities and differences in the data. Unlike with supervised learning methodologies, there is no corrective measure for unsupervised learning. The aspect of ground truth is missing, but the concept of optimization still exists.

Let's dive into the concepts of discriminative and generative models. In a generative model, the joint probability of inputs and output is learned. The distribution of the data is learned and is often the more generalized way of training a model. The models are capable of generating synthetic data points in the input space. On the other hand, discriminative models specialize in creating a mapping function from the input space to the output. Examples of generative models are Linear Discriminative Analysis, Naïve-Bayes, and Gaussian models.

Why are we introducing the generative models and discussing the ideologies of learning data distribution? Let's look back at the logic, which can help us with the super-resolution.

- Up-scaling images using the concept of nearest neighbors

- Bilinear interpolation/bicubic interpolation

- Fourier transforms

- Neural networks

We explore all these probable approaches in detail. But before this, let's explore the basic techniques used to up-scale a low-res image, starting with nearest-neighbor scaling. Figure 8-1a shows a basic image that can be resized to a bigger image (see Figure 8-1b), but keep in mind that the information in the image remains the same. It is just the representation that changes.

Figure 8-1a. *A 3x3 image*

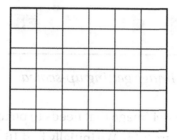

Figure 8-1b. *A 3x3 image extended to 6x6*

Up-Scaling Using the Nearest Neighbor Concept

Problems that demand faster resolution changes will require faster operations as well. We know that working with convolutional neural networks or anything close to neural networks is going to take a lot of computation, so it is time for some simple techniques. Up-scaling an image using the nearest neighbor concept is among the top contenders if we require a faster technique.

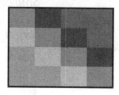

Figure 8-2a. *Sample image*

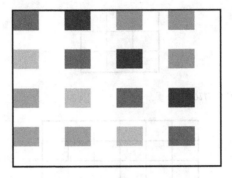

Figure 8-2b. *Sample Image getting up-scaled*

Figure 8-2a shows a 4x4 image that needs to be up-scaled to an 8x8 image, shown in Figure 8-2b. We initially had 16 pixels across the image and then, when it was stretched to 64 pixels, we were left with 48 vacancies that needed to be filled in. The concept of nearest neighbor can be understood from a straight line unit. Consider a straight number line, starting at 0 and ending at 4. If we divide it into four equal parts or in this case pixels, each part gets 25% of the information. Now, if the same line is stretched to have a length of 8, the units remain the same length, but the weight for each unit becomes 12.5%. However, the information carried in the image is the same.

The same concept can be used in nearest neighbor fulfillment of the vacancies using the following formula:

$$Source_X = Target_X \frac{Target_{width}}{Source_{width}}$$

$$Source_Y = Target_Y \frac{Target_{height}}{Source_{height}}$$

The formula provides us with the coordinate values for the enlarged image's pixels.

Understanding Bilinear Up-Scaling

To get to the concept of bilinear image up-scaling, we still have to cover linear interpolation. Interpolation suggests a one-dimensional extension of up-scaling. Consider a scenario where a straight line is marked by two values—x_1 and x_2—at two ends and is provided with the same values. If we have to interpolate a third value that lies between the ends, how do we proceed?

The algorithm suggests that we can use the concept of weighted averages to interpolate the unknown value. The weight can be found from the proportionate distance between the points x_1 and x_2. This logic can be used when we are resizing the image in two dimensions.

To get values for one coordinate in the (width, height) dimensions, we can work out the linear interpolation in each of the dimensions. That will essentially help with image resizing in two dimensions.

Moving on to the most awaited concepts in image up-scaling, which is neural networks. The methods of up-scaling may seem very rugged and cutthroat, with no finesse whatsoever. The logic to produce values using some tried-and-tested formulas was already there and was being used over and over again. The repetition can work wonders in some cases, but there are no changes or learning that is happening in the interim. Thus, we step into the learning section of neural networks. Before we write code and use the model architecture, we need to discuss the foundation blocks—VAE and GANs.

Variational Autoencoders

One of the most revolutionary improvements in the field of deep learning is the encoder-decoder architecture. Neural networks can retrieve the information present in an image and re-create the image from their understanding. Autoencoder architecture is a neural network architecture that can learn the patterns in the data and reduce it to smaller dimensions. These dimensions again can be used to re-create the images back to the original image. It's important to note that, theoretically, we can create a lossless architecture, with absolute re-creation of the image. In reality, such instances are rare.

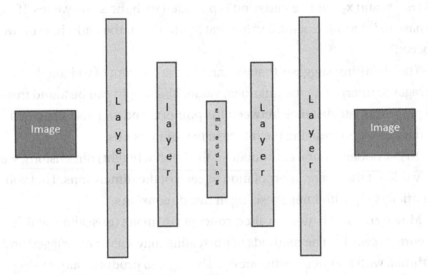

Figure 8-3. *Encoder-decoder architecture*

The neural network representational architecture is shown in Figure 8-3, and it demonstrates a situation where images are understood by the interconnected layers and converted to embedding. This embedding is the representation of the information from the images as per the established model architecture.

The initial part of the network, popularly called an *encoder,* learns the data distribution or the pattern in the input images. It not only has to understand itself, but the paired decoder architecture also needs to decode the embedding. Thus, the feature extraction and the understanding need to be such that the decoder can decipher the original image in question from the embedding with very minimal loss.

The decoder side, on the other hand, starts right after the embedding layer and tries to transform the embedding layer into the original image with minimum loss of information. This compression of information and then regeneration of the image by neural networks is the concept of an auto-encoder network.

When information is transmitted, the bandwidth of the transmission can impact the resolution of the images. Compression can help in passing the low-resolution images to the destination. Once the image reaches the destination, the decoder layers spring into action, up-scale, and resume the original image.

The concept of compression and decompressing can be further realized into up-scaling images. Now that we have explored the basics of the encoder-decoder architecture, will move on to an interesting concept, called *variational encoder*.

As we have seen so far, the traditional autoencoder architecture creates a latent space with the representational information from the input so that the output can be generated by the decoder network. But imagine that a single attribute is contributing to a discrete value, and it will be restricted to only one value when it is getting re-created. This restriction will not help the model generate something new from the distribution, but will just repeat. What if we want to get the representation in the latent space as a distribution instead of as a discrete value? We can do that, but there will be two different characteristics that come up now:

- Stochastic process

- Deterministic process

It is already established via multiple deep learning concepts, that whenever a network is getting trained, it needs to have a set of processes that can learn and adapt to losses. There will be a forward propagation to compute the loss (the difference between the actual and the output) given the parameters of the model. There will also be backward propagation, which will change the weights according to the loss incurred due to the differences in the expected output and the ground truth. So, essentially we can train a network only when it is deterministic. Variational autoencoders are shown in Figure 8-4.

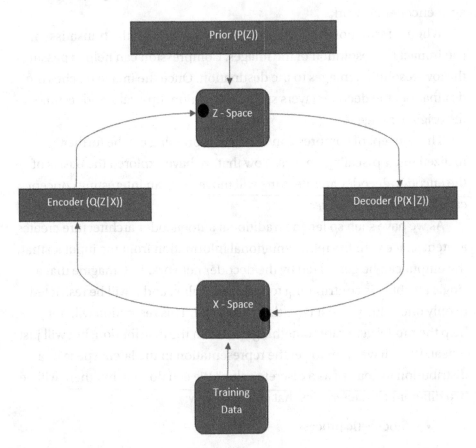

Figure 8-4. *VAE representation*

From the representational image shown in Figure 8-4, we can see how the variational autoencoder tries to map the data distribution to the latent space. The training data is used by the encoder network parameterized by φ to learn the stochastic mapping from the training data or X-space to the latent or Z-space.

The encoder or the inference model learns the pattern in the data. It can be proved that the empirical distribution of the X-space is complicated but the latent space is simple. The generative network parameterized by θ learns the distribution given by P(X|Z). The decoder part is learning from a prior distribution, which is usually a normal Gaussian distribution, and the deterministic process. To put forth the differences with an autoencoder, there are is an additional stochastic process that is added. Figure 8-5 shows the representation of the variational autoencoder network. It shows the stochastic nature being added to the existing autoencoder architecture.

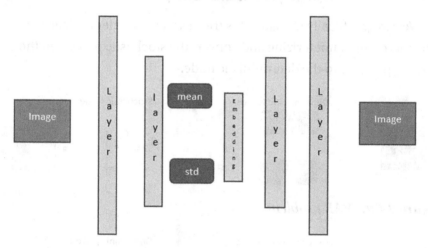

Figure 8-5. *VAE network representation*

So while earlier we were more focused on finding the vector of the latent space or the embedding of discrete values, now we will be finding a vector space of the mean and standard deviation.

The latent distribution is giving us the stochastic nature of the process. Eventually, we are going to have to backpropagate to train the model. To overcome this problem of training, we consider the mean to be a fixed vector. To keep the randomness as well as maintain the prior distribution infused in the model, we consider standard deviation to be a fixed vector affected by the random constants from the Gaussian prior distribution.

This sampling process is not as simple as it looks, given our loss function will be reconstruction loss and another regularization loss. We use a reparameterization trick, in which € is sampled from the prior Normal Gaussian distribution, and shifted by the mean of the latent distribution, then scaled by the standard deviation. The formula will be:

$$Z = mean + std * € ----- (i)$$

From the standard stochastic node, we get this equation:

$$Z = Q(Z|X) \text{ parameterized by } \varphi ---------- (ii)$$

We can graphically visualize this trick as well, in order to clear the concepts of reparameterizing and convert the stochastic process in the learning pathway to the deterministic node.

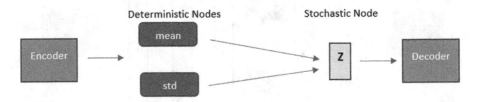

Figure 8-6a. *VAE problem*

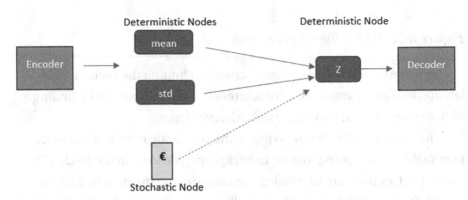

Figure 8-6b. *Reparameterization trick*

Figure 8-6a shows the problem arising from the backpropagation or essentially the model learning the latent space. Figure 8-6b shows the process of reparameterization, wherein the backpropagation can happen through the channels of solid arrow lines. The dotted arrow line shown is the stochastic process, which doesn't hinder the training process and does not take part in the backpropagation directly. It is not learning anything and no weights are getting adjusted per the loss function. It is interesting to note how the process type changes in the Z-space by the shift of the stochastic process away from the backpropagation path.

Equation (i) can be considered a rough estimation of Figure 8-6b and equation (ii) is an estimation of Figure 8-6a.

Thus we have established the concepts of variational autoencoders, which can have various uses. This small stochastic process can help generate similar images, drawn from the same probability distribution. It is helpful in image *regeneration* or image *generation,* and there will always be a need for both types. The sampling process enables the generator model or the decoder model to re-create images from the same distribution with subtle changes. In some cases, it helps in interpolating signals or images. This interpolation concept can be used to resize the images. Now that we covered variational autoencoders in brief, let's look at another generative form of algorithms, known as *generative adversarial networks,* before diving into the image resizing code.

Generative Adversarial Networks

Generative adversarial networks were introduced to deep learning by Ian Goodfellow in 2014. The network is capable of creating newer samples that are quite close to the original ones. They are also used extensively for style transfers in images.

This network is a combination of two models—the generator and the discriminator models. In combination, these models form a supervised form of learning.

- **Generator:** The model tries to generate samples based on a domain or a problem set. These are preferably samples out of a fixed distribution. The generator takes in random input (in most scenarios, a Gaussian distribution is used to help it with the input). During the training process, these random or meaningless points will be deemed to be coming out from the domain distribution. The generator should be able to generate the representation from the input data distribution. As we have seen earlier, the data distribution is complex and the encoder tries to map to a much simpler yet highly compressed information block. This space is often termed *latent space*, from which the generator block of the autoencoder generates the output. The models can understand the intricacies of the data distribution and create a representation, from which a sample is taken. This can and should be able to fool the discriminator or the classifier.

- **Discriminator:** Once the generator model creates fake samples that it thinks are very similar to the original data distribution, they are passed to the discriminator model to validate and classify. It is essentially a classification model. It has the job of classifying the images generated by the generator, fake or real. The classifier discriminates between real and fake images.

We have established that generators and discriminators have to be trained simultaneously. This is called a generative adversarial network since generative models and discriminative models are adversarial to each other. They are trying to better each other at a zero-sum game. One does not theoretically defeat the other. We have the generator network trying to create a fake image as realistically as possible such that the discriminator cannot identify it as fake. On the other hand, discriminator models are trying to train so that anything amiss in the image will be captured by it. In a perfect world, the generator eventually generates images that the discriminator can't identify as fake or real (50% fake/real). Eventually, the generator is removed from the network and used for other purposes.

The Model Code

We have discussed the basic concepts behind a generative adversarial network. This led to one of its many usages, super-resolution. It has various applications, among which style transfers, image generations, and super resolutions are few. The model that deals with super-resolution is SRGAN. One of its predecessors, called SRResNet, came up with good results in terms of SSIM and PSNR.

Let's look at the metrics that are usually determined in super-resolution problems:

- **Structural similarity index (SSIM):** The metric tries to put a value to the amount of degradation that happened due to changes from the source image to the target image. It checks the perceived similarity between sections of the image. It is based on the average and standard deviation of the chosen window.

- **Peak signal to noise ratio:** This is another important
  metric used to measure the reconstruction loss
  of an image or a changed image with the original
  image. It at best can be defined by the mean squared
  error computation. It can be formed by taking the
  logarithmic scale with a base of 10.

- **Mean opinion score (MOS):** This is defined by a single
  number on an ordinal scale, with a range of 1 to 5. 1
  is the lowest perceived quality and 5 is the highest
  perceived quality.

Now that we have looked at the metric to define and measure the
differences, let's look at the data on which we are going to develop
our code.

We will be using the DIV2K dataset, which has 1000 high-definition
images, with an 800-100-100 split in terms of train, validation, and test
data. This data can be downloaded from the source paper introduced in
CVPR 2017, at `https://data.vision.ee.ethz.ch/cvl/DIV2K/`.

The setup for the code needs to follow the application's standard
build. Normally this means that there needs to be a model file, a few utility
scripts, a training file, and a validation file. In a few scenarios, this model
needs to be an application hosted in a server, and there needs to be a setup
file as well. Taking this one step at a time, we can start with the model files.

Model Development

The codebase has a generative model block, the discriminative model
block, a residual block, and a content loss calculation block.

Imports

We will be using the Torch framework for the entire code block. If the development is done in a local environment, then we must make sure the Torch and its dependencies are installed in the environment and are working. Torch and TorchVision are the two important packages that need to be set up. If there is a provision for a GPU with CUDA cores, we should install the latest CUDA packages to help PyTorch utilize parallel GPU cores for computation. For the model script, we import the Torch- and TorchVision-related functions.

```python
import torch
import torch.nn as nn
import torch.nn.functional as F
import torchvision.models as models
from torch import Tensor
```

Next, we define the generator class to help regenerate the images.

```python
class Generator(nn.Module):
    ## defining the generator model
    ## extending the class
    ## initializing sequential - network expecting 64x3

    def __init__(self) -> None:
        super(Generator, self).__init__()

        self.convolutional_block1 = nn.Sequential(
            nn.Conv2d(3, 64, (9, 9), (1, 1), (4, 4)),
            nn.PReLU()
        )
```

```
## adding resnet conv block of 16

res_trunk = []
for _ in range(16):
    res_trunk.append(ResidualConvBlock(64))
self.res_trunk = nn.Sequential(*res_trunk)

self.convolutional_block2 = nn.Sequential(
    nn.Conv2d(64, 64, (3, 3), (1, 1), (1, 1),
    bias=False),
    nn.BatchNorm2d(64)
)

self.upsampling = nn.Sequential(
    nn.Conv2d(64, 256, (3, 3), (1, 1), (1, 1)),
    nn.PixelShuffle(2),
    nn.PReLU(),
    nn.Conv2d(64, 256, (3, 3), (1, 1), (1, 1)),
    nn.PixelShuffle(2),
    nn.PReLU()
)

self.convolutional_block3 = nn.Conv2d(64, 3, (9, 9),
(1, 1), (4, 4))

self._initialize_weights()

def forward(self, x: Tensor) -> Tensor:
    return self._forward_impl(x)

def _forward_impl(self, x: Tensor) -> Tensor:
    ## defining forward pass -> 3 convolutional blocks
    out1 = self.convolutional_block1(x)
    out = self.res_trunk(out1)
    out2 = self.convolutional_block2(out)
    output = out1 + out2
```

```
        output = self.upsampling(output)
        output = self.convolutional_block3(output)

        return output

    def _initialize_weights(self) -> None:
        ## initialize the weights
        ## adding provision for batch normalization
        for m in self.modules():
            if isinstance(m, nn.Conv2d):
                nn.init.kaiming_normal_(m.weight)
                if m.bias is not None:
                    nn.init.constant_(m.bias, 0)
                m.weight.data *= 0.1
            elif isinstance(m, nn.BatchNorm2d):
                nn.init.constant_(m.weight, 1)
                m.weight.data *= 0.1
```

This code snippet defines the class of the convolutional block that's capable of regenerating images. Importantly, this code block consists of three convolutional blocks and one up-sampling block. The first convolutional block is followed by a residual block, which serves as the trunk of the entire generator network. This is followed by a second convolutional block. The up-scaling block consists of a pair of convolutional layers followed by pixel shuffling. Eventually, the final convolutional block is added to generate the output. The block is provided with batch normalization layers and a combination of 3x3 convolution layers.

The forward pass helps build the sequential model in the function forward implementation. There is also another function to initialize the weights. Having covered the basic generator class, we will move to the next class of the discriminator.

The discriminator module extends the standard nn.module following eight layers of convolutions. They are using batch normalization after each layer to run deep. The model structure uses leaky ReLU for activation functions. The model ends with a torch.flatten layer, which helps it classify things.

```python
class Discriminator(nn.Module):
    ## defining discriminator
    def __init__(self) -> None:
        super(Discriminator, self).__init__()
        self.features = nn.Sequential(

            nn.Conv2d(3, 64, (3, 3), (1, 1), (1, 1), bias=True),
            nn.LeakyReLU(0.2, True),

            nn.Conv2d(64, 64, (3, 3), (2, 2), (1, 1), bias=False),
            nn.BatchNorm2d(64),
            nn.LeakyReLU(0.2, True),
            nn.Conv2d(64, 128, (3, 3), (1, 1), (1, 1),
            bias=False),
            nn.BatchNorm2d(128),
            nn.LeakyReLU(0.2, True),

            nn.Conv2d(128, 128, (3, 3), (2, 2), (1, 1),
            bias=False),
            nn.BatchNorm2d(128),
            nn.LeakyReLU(0.2, True),
            nn.Conv2d(128, 256, (3, 3), (1, 1), (1, 1),
            bias=False),
            nn.BatchNorm2d(256),
            nn.LeakyReLU(0.2, True),
```

```
        nn.Conv2d(256, 256, (3, 3), (2, 2), (1, 1),
        bias=False),
        nn.BatchNorm2d(256),
        nn.LeakyReLU(0.2, True),
        nn.Conv2d(256, 512, (3, 3), (1, 1), (1, 1),
        bias=False),
        nn.BatchNorm2d(512),
        nn.LeakyReLU(0.2, True),

        nn.Conv2d(512, 512, (3, 3), (2, 2), (1, 1),
        bias=False),
        nn.BatchNorm2d(512),
        nn.LeakyReLU(0.2, True)
    )

    self.classifier = nn.Sequential(
        nn.Linear(512 * 6 * 6, 1024),
        nn.LeakyReLU(0.2, True),
        nn.Linear(1024, 1),
        nn.Sigmoid()
    )

def forward(self, x: Tensor) -> Tensor:
    ## define forward pass
    output = self.features(x)
    output = torch.flatten(output, 1)
    output = self.classifier(output)

    return output
```

The model establishes the discriminator class in the architecture. Let's look at the ContentLoss class.

```python
class ContentLoss(nn.Module):

    ## defining content loss class
    ## feature extractors - till 36
    def __init__(self) -> None:
        super(ContentLoss, self).__init__()
    ## use pretrained VGG model to extract features
        vgg19_model = models.vgg19(pretrained=True,
        num_classes=1000).eval()

        self.feature_extractor = nn.Sequential(*list(vgg19_
        model.features.children())[:36])

        for parameters in self.feature_extractor.parameters():
            parameters.requires_grad = False

        self.register_buffer("std", torch.Tensor([0.229, 0.224,
        0.225]).view(1, 3, 1, 1))
        self.register_buffer("mean", torch.Tensor([0.485,
        0.456, 0.406]).view(1, 3, 1, 1))

    def forward(self, sr: Tensor, hr: Tensor) -> Tensor:
        hr = (hr - self.mean) / self.std
        sr = (sr - self.mean) / self.std

        mse_loss = F.mse_loss(self.feature_extractor(sr), self.
        feature_extractor(hr))

        return mse_loss
```

The class is using a pretrained VGG network to extract features to compute content loss. Following this, we look at a residual convolutional block.

```python
class ResidualConvBlock(nn.Module):
    ## get residual block

    def __init__(self, channels: int) -> None:
        super(ResidualConvBlock, self).__init__()
        self.rc_block = nn.Sequential(
            nn.Conv2d(channels, channels, (3, 3), (1, 1),
            (1, 1), bias=False),
            nn.BatchNorm2d(channels),
            nn.PReLU(),
            nn.Conv2d(channels, channels, (3, 3), (1, 1),
            (1, 1), bias=False),
            nn.BatchNorm2d(channels)
        )

    def forward(self, x: Tensor) -> Tensor:
        identity = x

        output = self.rc_block(x)
        output = output + identity

        return output
```

This concludes the model scripts. After this, we look at a few helper functions, starting with creating the dataset.

```python
def main():
    r""" Train and test """
    image_list = os.listdir(os.path.join("train", "input"))

    test_img_list = random.sample(image_list,
                                    int(len(image_list) / 10))
```

```
## Iterating through the test files

for test_img_file in test_img_list:
    filename = os.path.join("train", "input", test_
    img_file)
    logger.info(f"Process: `{filename}`.")

    shutil.move(os.path.join("train", "input", test_img_file),
                os.path.join("test", "input", test_img_file))
    shutil.move(os.path.join("train", "target", test_img_file),
                os.path.join("test", "target", test_img_file))
```

The function helps define the train test separation and locate it for
the training job to pursue. Another important function in line is the crop
function, which we can check next. It helps give back the cropped image.

```
def crop_image(img, crop_sizes: int):
    assert img.size[0] == img.size[1]
    crop_num = img.size[0] // crop_sizes

    box_list = []
    for width_index in range(0, crop_num):
        for height_index in range(0, crop_num):
            box_info = ( (height_index + 0)*crop_sizes,(width_
            index + 0) * crop_sizes,
                    (height_index + 1)*crop_sizes,(width_index +
                    1) * crop_sizes)
            box_list.append(box_info)

    cropped_images = [img.crop(box_info) for box_info in
    box_list]
    return cropped_images
```

One of the next functions that is important to deal with is the dataset
class. The dataset class provides batches of information to the training
function based on the configuration and availability.

```python
class BaseDataset(Dataset):
    ## base dataset class extending the dataset class
    from pytorch
    ## applies augmentation techniques such as random crop,
        rotation
    ## horizontal flip and tensor
    ## resizing and center crop is also used
    ## final conversion to tensor

    def __init__(self, dataroot: str, image_size: int, upscale_
    factor: int, mode: str) -> None:
        super(BaseDataset, self).__init__()
        self.filenames = [os.path.join(dataroot, x) for x in
        os.listdir(dataroot)]
        lr_img_size = (image_size // upscale_factor, image_size
        // upscale_factor)
        hr_img_size = (image_size, image_size)

        if mode == "train":
            self.hr_transforms = transforms.Compose([
                transforms.RandomCrop(hr_img_size),
                transforms.RandomRotation(90),
                transforms.RandomHorizontalFlip(0.5),
                transforms.ToTensor()
            ])
        else:
            self.hr_transforms = transforms.Compose([
                transforms.CenterCrop(hr_img_size),
                transforms.ToTensor()
            ])
        self.lr_transforms = transforms.Compose([
            transforms.ToPILImage(),
```

```
            transforms.Resize(lr_img_size, interpolation=IMode.
            BICUBIC),
            transforms.ToTensor()
        ])

    def __getitem__(self, index) -> Tuple[Tensor, Tensor]:
        hr = Image.open(self.filenames[index])
        temp_lr = self.lr_transforms(hr)
        temp_hr = self.hr_transforms(hr)
```

The dataset base class provides augmentation functions such as random crop, center crop, random rotation, horizontal flip, and resizing. Eventually, it converts the data to tensors for the PyTorch framework. It also has length and get item functions.

After all the important functions needed for development, we come down to the training sequence. The training sequence trains the generator. The code block follows:

```
def train_generator(train_dataloader, epochs) -> None:
    ## starting with train generator
    ## defining data loader
    ## defining the loss function
    batch_count = len(train_dataloader)
    ## start training for generator block
    generator.train()

    for index, (lr, hr) in enumerate(train_dataloader):
        ## getting hr to cuda or cpu
        hr = hr.to(device)
        ## getting lr to cuda or cpu
        lr = lr.to(device)
```

```
## initializing to zero grad for generator to avoid
gradient accumulation
## Accumulation is only suggested in case of time
base model
generator.zero_grad()

sr = generator(lr)
## defining pixel loss
pixel_losses = pixel_criterion(sr, hr)
## get step function from optimizers
pixel_losses.backward()
## adam optimizers for generator
p_optimizer.step()

iteration = index + epochs * batch_count + 1
writer.add_scalar(" computing train generator Loss",
pixel_losses.item(), iteration)
```

Similarly, the training for the adversarial block is as follows.

```
def train_adversarial(train_dataloader, epoch) -> None:
    ## for training adversarial network

    batches = len(train_dataloader)
    ## training discriminator and generator
    discriminator.train()
    generator.train()

    for index, (lr, hr) in enumerate(train_dataloader):
        hr = hr.to(device)
        lr = lr.to(device)
```

```
label_size = lr.size(0)
fake_label = torch.full([label_size, 1], 0.0, dtype=lr.
dtype, device=device)
real_label = torch.full([label_size, 1], 1.0, dtype=lr.
dtype, device=device)

## initializing zero grad since we want to avoid grad
accumulation
discriminator.zero_grad()

output_dis = discriminator(hr)
dis_loss_hr = adversarial_criterion(output_dis, real_label)
dis_loss_hr.backward()
dis_hr = output_dis.mean().item()

sr = generator(lr)

output_dis = discriminator(sr.detach())
dis_loss_sr = adversarial_criterion(output_dis, fake_label)
dis_loss_sr.backward()
dis_sr1 = output_dis.mean().item()

dis_loss = dis_loss_hr + dis_loss_sr
d_optimizer.step()

generator.zero_grad()

output = discriminator(sr)

pixel_loss = pixel_weight * pixel_criterion(sr,
hr.detach())
perceptual_loss = content_weight * content_
criterion(sr, hr.detach())
adversarial_loss = adversarial_weight * adversarial_
criterion(output, real_label)
```

```
gen_loss = pixel_loss + perceptual_loss +
adversarial_loss
gen_loss.backward()
g_optimizer.step()
dis_sr2 = output.mean().item()

iteration = index + epoch * batches + 1
writer.add_scalar("Train_Adversarial/D_Loss", dis_loss.
item(), iteration)
writer.add_scalar("Train_Adversarial/G_Loss", gen_loss.
item(), iteration)
writer.add_scalar("Train_Adversarial/D_HR", dis_hr,
iteration)
writer.add_scalar("Train_Adversarial/D_SR1", dis_sr1,
iteration)
writer.add_scalar("Train_Adversarial/D_SR2", dis_sr2,
iteration)
```

Eventually, we will work on a validation block, putting the generator and adversarial network together.

The following code puts everything together into the main function and runs the entire training sequence:

```
def main() -> None:

    ## creating directories
    ## making up training and validation datasets loc
    ## check for training
    ## check for resuming training if an opportunity
    if not os.path.exists(exp_dir1):
        os.makedirs(exp_dir1)
    if not os.path.exists(exp_dir2):
        os.makedirs(exp_dir2)
```

```
    train_dataset = BaseDataset(train_dir, image_size, upscale_
    factor, "train")
    train_dataloader = DataLoader(train_dataset, batch_size,
    True, pin_memory=True)

    valid_dataset = BaseDataset(valid_dir, image_size, upscale_
    factor, "valid")
    valid_dataloader = DataLoader(valid_dataset, batch_size,
    False, pin_memory=True)

    if resume:
        ## for resuming training
        if resume_p_weight != "":
            generator.load_state_dict(torch.load(resume_p_
            weight))
        else:
            discriminator.load_state_dict(torch.load(resume_d_
            weight))
            generator.load_state_dict(torch.load(resume_g_
            weight))

    best_psnr_val = 0.0

    for epoch in range(start_p_epoch, p_epochs):

        train_generator(train_dataloader, epoch)

        psnr_val = validate(valid_dataloader, epoch,
        "generator")

        best_condition = psnr_val > best_psnr_val
        best_psnr_val = max(psnr_val, best_psnr_val)
```

```python
    torch.save(generator.state_dict(), os.path.join(exp_
    dir1, f"p_epoch{epoch + 1}.pth"))
    if best_condition:
        torch.save(generator.state_dict(), os.path.
        join(exp_dir2, "p-best.pth"))

## saving best model
torch.save(generator.state_dict(), os.path.join(exp_dir2,
"p-last.pth"))

best_psnr_val = 0.0

generator.load_state_dict(torch.load(os.path.join(exp_dir2,
"p-best.pth")))

for epoch in range(start_epoch, epochs):

    train_adversarial(train_dataloader, epoch)

    psnr_val = validate(valid_dataloader, epoch, "adversarial")

    best_condition = psnr_val > best_psnr_val
    best_psnr_val = max(psnr_val, best_psnr_val)

    torch.save(discriminator.state_dict(), os.path.
    join(exp_dir1, f"d_epoch{epoch + 1}.pth"))
    torch.save(generator.state_dict(), os.path.join(exp_
    dir1, f"g_epoch{epoch + 1}.pth"))
    if best_condition:
        torch.save(discriminator.state_dict(), os.path.
        join(exp_dir2, "d-best.pth"))
        torch.save(generator.state_dict(), os.path.
        join(exp_dir2, "g-best.pth"))

    d_scheduler.step()
    g_scheduler.step()
```

```
torch.save(discriminator.state_dict(), os.path.join(exp_
dir2, "d-last.pth"))
torch.save(generator.state_dict(), os.path.join(exp_dir2,
"g-last.pth"))
```

With this, we conclude the code and can look at how to run it. The
code block should look like Figure 8-7. Once that is done, we can move to
the next section, which covers running the application.

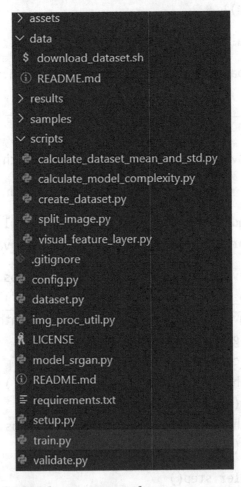

Figure 8-7. *Code development template*

Running the Application

To run the application, we need to start by downloading the dataset to the proper directory or map the data directory to the training function via the configuration script. The configuration script is important since it binds all the scripts and locations together. It helps the application understand what is needed.

To download the data, we can access the download script with bash.

```
! bash ./data/download_dataset.sh
```

Following the installation, we simply run the training script.

```
! python train.py
```

Once the generator training is done, the adversarial training will start. We can get a quick look at how the epochs might look.

```
Train Epoch[0016/0020](00010/00050) Loss: 0.008974.
Train Epoch[0016/0020](00020/00050) Loss: 0.009684.
Train Epoch[0016/0020](00030/00050) Loss: 0.004455.
Train Epoch[0016/0020](00040/00050) Loss: 0.008851.
Train Epoch[0016/0020](00050/00050) Loss: 0.008883.
Valid stage: generator Epoch[0016] avg PSNR: 21.19.

Train Epoch[0017/0020](00010/00050) Loss: 0.005397.
Train Epoch[0017/0020](00020/00050) Loss: 0.006351.
Train Epoch[0017/0020](00030/00050) Loss: 0.007704.
Train Epoch[0017/0020](00040/00050) Loss: 0.007926.
Train Epoch[0017/0020](00050/00050) Loss: 0.005559.
Valid stage: generator Epoch[0017] avg PSNR: 21.37.

Train Epoch[0018/0020](00010/00050) Loss: 0.006054.
Train Epoch[0018/0020](00020/00050) Loss: 0.008028.
Train Epoch[0018/0020](00030/00050) Loss: 0.006164.
```

Train Epoch[0018/0020](00040/00050) Loss: 0.006737.
Train Epoch[0018/0020](00050/00050) Loss: 0.007716.
Valid stage: generator Epoch[0018] avg PSNR: 21.36.

Train Epoch[0019/0020](00010/00050) Loss: 0.009527.
Train Epoch[0019/0020](00020/00050) Loss: 0.004672.
Train Epoch[0019/0020](00030/00050) Loss: 0.004574.
Train Epoch[0019/0020](00040/00050) Loss: 0.005196.
Train Epoch[0019/0020](00050/00050) Loss: 0.007712.
Valid stage: generator Epoch[0019] avg PSNR: 21.64.

Train Epoch[0020/0020](00010/00050) Loss: 0.006843.
Train Epoch[0020/0020](00020/00050) Loss: 0.007701.
Train Epoch[0020/0020](00030/00050) Loss: 0.005366.
Train Epoch[0020/0020](00040/00050) Loss: 0.004797.
Train Epoch[0020/0020](00050/00050) Loss: 0.008607.
Valid stage: generator Epoch[0020] avg PSNR: 21.53.

Train stage: adversarial Epoch[0001/0005](00010/00050) D Loss:
0.051520 G Loss: 0.574723 D(HR): 0.970196 D(SR1)/D(SR2):
0.019971/0.003046.
Train stage: adversarial Epoch[0001/0005](00020/00050) D Loss:
0.001356 G Loss: 0.528222 D(HR): 0.998656 D(SR1)/D(SR2):
0.000007/0.000005.
Train stage: adversarial Epoch[0001/0005](00030/00050) D Loss:
0.004768 G Loss: 0.574079 D(HR): 0.999959 D(SR1)/D(SR2):
0.004646/0.000619.
Train stage: adversarial Epoch[0001/0005](00040/00050) D Loss:
0.000339 G Loss: 0.557449 D(HR): 0.999820 D(SR1)/D(SR2):
0.000159/0.000527.
Train stage: adversarial Epoch[0001/0005](00050/00050) D Loss:
0.009615 G Loss: 0.531170 D(HR): 0.990858 D(SR1)/D(SR2):
0.000000/0.000000.
Valid stage: adversarial Epoch[0001] avg PSNR: 11.47.

Train stage: adversarial Epoch[0002/0005](00010/00050) D Loss: 0.000002 G Loss: 0.488294 D(HR): 0.999998 D(SR1)/D(SR2): 0.000000/0.000000.

Train stage: adversarial Epoch[0002/0005](00020/00050) D Loss: 0.114398 G Loss: 0.568630 D(HR): 0.947419 D(SR1)/D(SR2): 0.000000/0.000000.

Train stage: adversarial Epoch[0002/0005](00030/00050) D Loss: 3.704494 G Loss: 0.580344 D(HR): 0.230086 D(SR1)/D(SR2): 0.000000/0.000000.

Train stage: adversarial Epoch[0002/0005](00040/00050) D Loss: 0.000804 G Loss: 0.557581 D(HR): 0.999662 D(SR1)/D(SR2): 0.000464/0.000324.

Train stage: adversarial Epoch[0002/0005](00050/00050) D Loss: 0.001132 G Loss: 0.459117 D(HR): 0.999191 D(SR1)/D(SR2): 0.000317/0.000301.

Valid stage: adversarial Epoch[0002] avg PSNR: 12.48.

Train stage: adversarial Epoch[0003/0005](00010/00050) D Loss: 0.000187 G Loss: 0.488436 D(HR): 0.999847 D(SR1)/D(SR2): 0.000033/0.000032.

Train stage: adversarial Epoch[0003/0005](00020/00050) D Loss: 0.001537 G Loss: 0.444651 D(HR): 0.999899 D(SR1)/D(SR2): 0.001425/0.001385.

Train stage: adversarial Epoch[0003/0005](00030/00050) D Loss: 0.000169 G Loss: 0.493448 D(HR): 0.999877 D(SR1)/D(SR2): 0.000046/0.000041.

Train stage: adversarial Epoch[0003/0005](00040/00050) D Loss: 0.000285 G Loss: 0.465992 D(HR): 0.999925 D(SR1)/D(SR2): 0.000210/0.000202.

Train stage: adversarial Epoch[0003/0005](00050/00050) D Loss: 0.000720 G Loss: 0.567912 D(HR): 0.999978 D(SR1)/D(SR2): 0.000695/0.000668.

Valid stage: adversarial Epoch[0003] avg PSNR: 13.09.

```
Train stage: adversarial Epoch[0004/0005](00010/00050) D Loss:
0.000293 G Loss: 0.479247 D(HR): 0.999786 D(SR1)/D(SR2):
0.000079/0.000076.
Train stage: adversarial Epoch[0004/0005](00020/00050) D Loss:
0.000064 G Loss: 0.492225 D(HR): 0.999978 D(SR1)/D(SR2):
0.000042/0.000041.
Train stage: adversarial Epoch[0004/0005](00030/00050) D Loss:
0.000030 G Loss: 0.444387 D(HR): 0.999984 D(SR1)/D(SR2):
0.000014/0.000014.
Train stage: adversarial Epoch[0004/0005](00040/00050) D Loss:
0.000108 G Loss: 0.387137 D(HR): 0.999918 D(SR1)/D(SR2):
0.000025/0.000025.
Train stage: adversarial Epoch[0004/0005](00050/00050) D Loss:
0.000224 G Loss: 0.513328 D(HR): 0.999825 D(SR1)/D(SR2):
0.000049/0.000048.
Valid stage: adversarial Epoch[0004] avg PSNR: 13.29.
```

On this training set, we use configurable epochs and other training parameters, which are all available in the config file. Once the model is ready for download, we can use it to up-scale the images four times. We can configure the up-scaling factors as we train. With this, our training process is concluded.

Summary

This chapter started with the problems related to up-scaling images and discussed how to proceed with up-scaling. We discussed the advantages of the various methods and the modeling techniques currently available. State-of-the-art algorithms like SRGAN were discussed and implemented.

We also went through the training process set up the project. This chapter discussed how we can use convolutional models to up-scale images by some factor in combination with generative models. Super-resolution is a growing field and is widely used, such as to detect license plates from a traffic camera or to enhance old photographs. This is a very important field in computer vision and holds years of research.

In the upcoming chapter, we move on from the concepts of still images to moving images, also known as video.

CHAPTER 9

Video Analytics

The machine learning journey started from structured data long ago to the process of extracting meaningful predictions. As data grew, machine learning started exploring other data types as well. Today, there is no limit to the types of data that can be processed.

From structured data, we started analyzing text data. We started understanding text and making predictions using features in the text. Then we jumped to images as well. Even though this process was challenging at times, thanks to the advancement in processing power of GPUs and TPUs, things started to fall in place.

Then comes audio processing. This involves processing the audio using frequency or converting the audio into text and then making predictions.

The combination of all these concepts is called *video analytics*.

The amount of video data available out there is enormous. The world is creating video content every second. The entertainment and sports industries run on videos, and security cameras are capturing every movement. As shown in Figure 9-1, YouTube alone has more than two billion users. Think about the amount of data being generated from videos. As we see more data, more problems will surface and AI will be leveraged to solve these problems.

© Akshay Kulkarni, Adarsha Shivananda, and Nitin Ranjan Sharma 2022
A. Kulkarni et al., *Computer Vision Projects with PyTorch*,
https://doi.org/10.1007/978-1-4842-8273-1_9

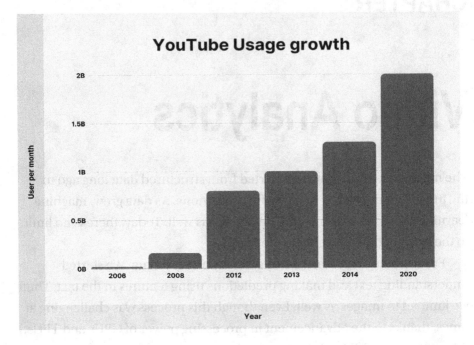

Figure 9-1. YouTube

Video content analysis or video content analytics, also known as video analysis or video analytics, is the capability of automatically analyzing video to detect and determine temporal and spatial events.

Problem Statement

Before we consider the problem statement, let's list a few of the possible data sources for video analytics.

- YouTube

- Social media

- Sports

- Entertainment industry

- Security cameras

- Teaching platforms

- Mobile recordings

The need for video processing is critical at this point in time. With the amount of data, manually auditing video footage created by various sources like surveillance cameras and other recordings is impossible.

Every industry has problems that can be solved using videos. Let's discuss a few of the uses here.

- Count people at malls, retail shops, etc.

- Personal demographics identification, such as age and gender, in videos

- Inventory planning for shelf replenishment alerts

- Security and surveillance at shops using motion detection

- Parking lot use cases

- Face recognition

- Behavior detection

- Person tracking

- Crowd detection

- People count/people presence

- Time management

- Zone management and analysis/boundary detection

- Traffic controlling systems

- Security/surveillance

- Motion detection

- Queue management

- At-home monitoring

- Automatic license plate recognition

- Traffic monitoring

- Vehicle counting

- Sports analytics, as shown in Figure 9-2

Figure 9-2. *Sports video analysis*

We have picked a few use cases from this list and will implement them. We are going to implement the following use cases using PyTorch.

- Count the number of customers at a retail outlet

- Identify the hotspot in retail stores

- Manage security and surveillance using motion detection

- Identify demographics (age and gender)

Approach

We will be using the video as input to our algorithms. The video basically has two components:

- Visuals or series of images

- Audio

We need to extract these sub-components from the video to process it and solve the use cases. We are only interested in the first component for this project. Once we extract a series of images based on the use cases, we can use algorithms or pretrained models. Figure 9-3 shows the solution approach for video analytics.

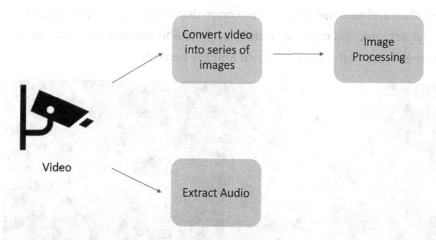

Figure 9-3. *Solution approach*

The last section, which is image processing, can include various tasks related to video analytics.

- **Image classification:** Classify the images extracted from the video. For example, identify the gender of the human being in the video.

301

- **Object detection:** Detect the object in an image. For example, detect cars in the parking lot.

- **Object tracking:** Once the objects are detected, identify the movement of the object.

- **Segmentation:** Generate the bounding box to identify the various objects in the image.

Implementation

Let's see how to implement each of these use cases. We need to first install the clone libraries. We are using a pretrained model based on the SFNet architecture. It has an encoder-decoder based convolutional neural network and dual-path multi-scale fusion networks with attention. Figure 9-4 shows the sample crowd and the heat map.

Figure 9-4. Crowd and heat map

We are also using the FaceLib library for face detection, see we can try to predict age and gender. And for image processing and any other manipulation, let's use OpenCV.

```
# Installing required packages

!pip install git+https://github.com/sajjjadayobi/FaceLib.git
!git clone https://github.com/Pongpisit-Thanasutives/
Variations-of-SFANet-for-Crowd-Counting #only model

# Importing Packages
import cv2
from PIL import Image
import pandas as pd
import numpy as np
%pylab inline
import matplotlib.pyplot as plt
import matplotlib.image as mpimg
import glob

#import torch
import torch
from torchvision import transforms

#import model
import os
os.chdir('/content/Variations-of-SFANet-for-Crowd-Counting')
from models import M_SFANet_UCF_QNRF

#import facelib
from facelib import FaceDetector, AgeGenderEstimator
```

Data

We use a couple of videos from YouTube. Download these videos and keep them local.

1.   Here is a video link from a retail store.
`https://www.youtube.com/watch?v=KMJS66jBtVQ`

2.   Here is the video link from a parking lot.
`https://www.youtube.com/watch?v=eE2ME4BtXrk`

Download those videos and then we will upload them to Google Colab to perform further analysis.

Uploading the Required Videos to Google Colab

For the next step, we need to upload the videos to Google Colab.

```
# Uploading the videos
from google.colab import files

#upload
files.upload()
files.upload()
```

We also need to upload a model weights file, which will be used later. Download it from the following link: `https://drive.google.com/file/d/1oaXIBVg-dgyqRvEXsYDiJh5GNzP35vO-/view`

Upload this model's weights file to Colab:

```
print('Upload model weights')
files.upload()
```

Convert the Video to a Series of Images

As discussed, video analytics is all about generating frames from video. The rest of the steps are the same as when image processing.

Let's build a function that generates images as we input the video.

```python
#function to generates images

def video_to_image(path, folder):
    global exp_fld

    # importing video
    vidcap=cv2.VideoCapture(path)
    exp_fld=folder

    # error handling
    try:
        if not os.path.exists(exp_fld):
            os.makedirs(exp_fld)
    except OSError:
        print ('Error: Creating directory of data')

    Count = 0
    sec = 0
    frameRate = 1 # secs of the video

    while(True):
        vidcap.set(cv2.CAP_PROP_POS_MSEC,sec*1000)
        hasFrames,image = vidcap.read()
        sec = sec + frameRate
        sec = round(sec, 2)

        # Exporting the image
        if hasFrames:
            name='./' + exp_fld +'/frame'+str(Count) + '.jpg'
            cv2.imwrite(name, image)        # save frame as
                                            JPG file

            Count +=1
        else:
            break
    return print("Image Exported")
```

Image Extraction

Now, let's generate the frames for the videos using the function we created previously. Let's do it for both videos.

```
#setting the path
os.chdir('/content')
```

```
# Extracting Images and Storing for a first video
video_to_image('HD CCTV Camera video 3MP 4MP iProx CCTV
HDCCTVCameras.net retail store.mp4', 'crowd')
```

```
# Extracting images for parking lot video
video_to_image('AI Security Camera with IR Night Vision (Bullet
IP Camera).mp4', 'movement')
```

```
Image Exported
Image Exported
```

Data Preparation

Now, let's run some quick data-preparation steps. As compared to structured data, preparing and cleaning data is easier when doing image processing. The key step here is to resize the images. The original images can be any size or pixels. But the models are always trained on a certain size. So we need to resize the input images to match the model.

Here is the function that resizes the images.

```
# Resizing the image
def img_re_sizing(dnst_mp, image):

    #Normalizing
    dnst_mp = 255*dnst_mp/np.max(dnst_mp)
    dnst_mp= dnst_mp[0][0]
    image= image[0]
```

```
#empty image
result_img = np.zeros((dnst_mp.shape[0]*2, dnst_
mp.shape[1]*2))

#iterate for each image
for i in range(result_img.shape[0]):
    for j in range(result_img.shape[1]):
        result_img[i][j] = dnst_mp[int(i / 2)]
        [int(j / 2)] / 4
result_img  = result_img.astype(np.uint8, copy=False)

#output
return result_img
```

Now that we have completed video ingestion, image extraction, and resizing, let's solve the use cases.

Let's start by counting the customers and generating heat maps. We need to count the number of customers at this retail outlet (see Figure 9-5).

Figure 9-5. *Retail store*

The revenue of a store depends on how many people visit it. As evident, the bigger the crowd, the greater the probability of purchases. That's why the store wants to understand how many people visit, so it can forecast sales. Many decisions are made from this information, like the size of the store, how to plan inventory, etc.

Identify the Hotspots in a Retail Store

It's very important to place inventory in a store so that it maximizes purchases. We can leverage hotspots within stores to make that decision. We can also identify trending products using this information. We can get this kind of information using the heat map concept. It generates spots where customers spend most of their time, which helps determine the demand for the product in that spot. Figure 9-6 shows a sample heat map.

Figure 9-6. Heat map

Let's write a few functions to generate a heat map and count the people. The following function will take images and generate the heat map.

```python
# function to get a heatmap

def generate_dstys_map(o, dsty, cc, image_location):

    #define the fgr_imure
    fgr_im=plt.fgr_imure()

    #define size
    col = 2
    rws = 1
    X = o

    #sum
    add = int(np.sum(dsty))

    dsty = image_re_sizing(dsty, o)

    # Adding original image and new generated heatmap image
    for i in range(1, col*rws +1):

        # generate original image
        if i == 1:
            image = X
            fgr_im.add_subplot(rws, col, i)
            #setting axis
            plt.gca().set_axis_off()
            plt.margins(0,0)
                #locator
            plt.gca().xaxis.set_major_locator(plt.
            NullLocator())
            plt.gca().yaxis.set_major_locator(plt.
            NullLocator())
                #adjusting subplots
            plt.subplots_adjust(top = 1, bottom = 0, right = 1,
            left = 0, hspace = 0, wspace = 0)
```

```
            #show image
        plt.imshow(image)

    # generate dstys image
    if i == 2:
        image = dsty
        fgr_im.add_subplot(rws, col, i)
            #setting axis
        plt.gca().set_axis_off()
        plt.margins(0,0)
            #locator
        plt.gca().xaxis.set_major_locator(plt.
        NullLocator())
        plt.gca().yaxis.set_major_locator(plt.
        NullLocator())
            #adjusting subplots
        plt.subplots_adjust(top = 1, bottom = 0, right = 1,
        left = 0, hspace = 0, wspace = 0)
            #adding count
        plt.text(1, 80, 'M-SegNet* Est: '+str(add)+',
        Gt:'+str(cc), fontsize=7, weight="bold",
        color = 'w')
            #show image
        plt.imshow(image)#, cmap=CM.jet)

#image_nm with location
image_nm = image_location.split('/')[-1]
image_nm = image_nm.replace('.jpg', '_heatpmap.png')

#saving the image
plt.savefgr_im(image_location.split(image_nm)
[0]+'seg_'+image_nm, transparent=True, bbox_inches='tight',
pad_inches=0.0, dpi=200)
```

310

The following function will take images as input and count the people. It also triggers the heat map function that we created just now. Finally, it outputs the count of people, the image density, and a density map.

```
# Getting count of peoples
def get_count_people(image):

    # Simple preprocessing.
    trans = transforms.Compose([transforms.ToTensor(),
                                transforms.Normalize([0.485,
                                0.456, 0.406], [0.229,
                                0.224, 0.225])
                                ])

    # sample image with height and width
    img = Image.open(image).convert('RGB')
    height, width = img.size[1], img.size[0]
    height = round(height / 16) * 16
    width = round(width / 16) * 16

    # re size the image
    img_den = cv2.resize(np.array(img), (width,height), cv2.
    INTER_CUBIC)

    #transform
    img = trans(Image.fromarray(img_den))[None, :]

    #lets define the model
    model = M_SFANet_UCF_QNRF.Model()

    #load the model
    model.load_state_dict(torch.load('/content/best_M-SFANet_
    UCF_QNRF.pth', map_location = torch.device('cpu')))

    # Evaluating the model
    model.eval()
```

```
    dnst_mp = model(img)

    # final count
    count = torch.sum(dnst_mp).item()

    #retun count, density and map
    return count,img_den,dnst_mp
```

At this point, we have created the functions to count and generate a heat map. Next, let's use these functions on the images we extracted from the video.

Before that, let's import the images.

Importing Images

```
# Getting all the images from the path
image_location = []
path_sets = ['/content/crowd']

#loading all the images
for path in path_sets:
    for img_path in glob.glob(os.path.join(path, '*.jpg')):
        image_location.append(img_path)
image_location[:3]

['/content/crowd/frame91.jpg',
 '/content/crowd/frame12.jpg',
 '/content/crowd/frame26.jpg']
```

Getting Crowd Counts

Let's iterate through every image we imported through the functions created previously. Finally, we append the count and image ID to create a dataframe that will act as the output.

```
# Getting the count of people by image

#define empty list
list_df = []

#loop through each image
for i in image_location :
    count, img_den, dnst_mp = get_count_people(i)
    generate_dens_map(img_den, dnst_mp.cpu().detach().
    numpy(), 0, i)
    list_df = list_df + [[i,count]]

#create the dataframe with image id and count
df = pd.DataFrame(list_df,columns=['image','count'])

#sort and show
df.sort_values(['image']).head()
```

Figure 9-7a shows the output from the model. It gives the count of the customer at each frame. This gives us an idea of how many customers are visiting on a daily basis. Let's do some basic statistics on this prediction.

```
Df.describe()
```

	image	count
78	/content/crowd/frame0.jpg	17.489319
71	/content/crowd/frame1.jpg	17.444080
13	/content/crowd/frame10.jpg	18.288996
65	/content/crowd/frame11.jpg	17.052418
1	/content/crowd/frame12.jpg	18.255342

Figure 9-7a. *The output*

```
import seaborn as sns
sns.histplot(data=df['count'])
```

As we can observe from Figure 9-7b and 9-7c, on average, 15 customers are present in the store at any one time.

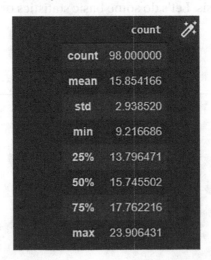

Figure 9-7b. *The output*

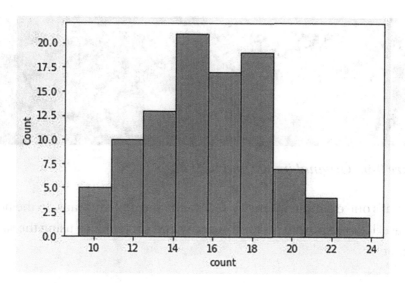

Figure 9-7c. *The output*

Let's look at a few of the heat maps that were generated as well.
Figures 9-8 and 9-9 show the heat map at given time instances. On the left
side, we have the original image, and on the right side we have the heat
map. The white and yellow patches indicate the customer density in that
particular area. Using this information, we can easily identify the hotspots
in the store.

Figure 9-8. *Original image and heat map*

Figure 9-9. *Original image and heat map*

That's one example where we saw count and density maps. In the next use case, let's see how to perform security and surveillance using the same concept.

Security and Surveillance

In this century, security has become a big concern. Video surveillance has been around for ages, but how it's being done is changing. There is room for a lot of automation. With AI, that is becoming reality. See Figure 9-10.

Figure 9-10. *Manual surveillance*

We can analyze if there is any movement in a restricted or sensitive area by detecting a person moving using video processing. This way, a live person doesn't have to keep looking at security footage 24*7. If there is any movement at all, the system can alert responsible security.

Models can be trained to predict an adverse event. Continuously monitoring for events using such software will save many personnel hours as well as decrease security risks.

Let's look at the second video we uploaded and detect movement in it.

```
# Getting all the images from the path
image_location = []
path_sets = ['/content/movement']

#loading all images for parking lot video
for path in path_sets:
```

```
    for img_path in glob.glob(os.path.join(path, '*.jpg')):
        image_location.append(img_path)

image_location[:3]


# Getting the count of people by image
list_df = []
```

Now that all the images are loaded, the next step is to iterate through every image and see if there are human beings in the no-entry zone.

Let's capture this presence of human beings in a new column of a dataframe, called movement.

```
#checking if there is any movement for each frame in video
for i in image_location :
    count, img_den, dnst_mp = get_count_people(i)
    generate_dens_map(img_den, dnst_mp.cpu().detach().
    numpy(), 0, i)
    list_df = list_df + [[i,count]]


#saving the data into dataframe
detected_df = pd.DataFrame(list_df,columns=['image','count'])
detected_df['movement'] = np.where(detected_df['count'] > 3
,'yes','no')
detected_df.filter(items = [45,56,53], axis=0)
```

Figure 9-11 shows the output. As we can observe, frame33 reports yes in the movement column, which means there is the presence of humans at that particular second.

	image	count	movement
45	/content/movement/frame31.jpg	2.973071	no
56	/content/movement/frame46.jpg	2.144534	no
53	/content/movement/frame33.jpg	3.724376	yes

Figure 9-11. *The output*

As soon as the movement variable becomes yes, we can alert the system to trigger the respective authorities.

Let's look at a few examples where we observe some movement.

```
#print the image where movement = 'yes'
image = '/content/movement/frame25.jpg'
Image.open(image)

image = '/content/movement/frame27.jpg'
Image.open(image)
```

Figures 9-12 and 9-13 show images with human beings in the frame and movement is being detected successfully. In the next section, we see how to detect age/gender in a sample video.

Figure 9-12. *The output*

Figure 9-13. *The output*

Identify the Demographics (Age and Gender)

Most marketing activities depend on the demographics of the target customers. If we can determine the age and gender from these real-time videos from malls, we can use this information to target appropriate marketing activities.

Let's try to detect age and gender using the images we extracted earlier.

```
# Importing functions
face_detector = FaceDetector()
age_gender_detector = AgeGenderEstimator()

# Reading image
img = '/content/movement/frame0.jpg'
image = cv2.imread(img, cv2.IMREAD_UNCHANGED)
Image.open(img)
```

Figure 9-14 shows the sample image for which gender and age will be detected.

```
#checking the gender

faces, boxes, scores, landmarks = face_detector.detect_
align(image)
genders, ages = age_gender_detector.detect(faces)
print(genders, ages)
```

['Male'] [32]

Figure 9-14. *Input image*

The model predicted the age and gender of the person in this image:

- Age is 32

- Gender is male

Summary

We explored various use cases of video analytics and picked four use cases to implement. The solution approach was discussed along with the libraries used to achieve the solution.

A key aspect is converting videos into images and then performing traditional image processing on that information. The main challenge is that videos generate a huge number of images, and processing them takes time and resources. In many cases, everything needs to happen in real-time to make full use of prediction.

We discussed a few fundamental use cases, but there is an ocean of use cases to explore, learn, and implement. There are also many technical challenges to be solved that are in the research phase. Now that we have learned and built some computer vision models for different applications, the next chapter dives deep into the output of computer vision models and discusses the explainable AI for computer vision.

CHAPTER 10

Explainable AI for Computer Vision

Most machine learning and deep learning models lack a way of explaining and interpreting results. Due to the dynamic nature of deep learning models and increasing state-of-the-art models, the current model evaluation is based on accuracy scores. This makes machine learning and deep learning black-box models. This leads to lack of confidence in applying the model and lack of trust of the generated results. There are multiple libraries that help us explain models of structured data like SHAP and LIME. This chapter explains computer vision model outputs.

Here are some of the white-box algorithms proposed in the recent years for computer vision:

- CAM

- Grad-CAM

- Grad-CAM++

- Layer-wise relevance propagation (LRP)

- SmoothGRAD

- RISE

- Neural based decision trees (NBDT)

© Akshay Kulkarni, Adarsha Shivananda, and Nitin Ranjan Sharma 2022
A. Kulkarni et al., *Computer Vision Projects with PyTorch*,
https://doi.org/10.1007/978-1-4842-8273-1_10

This chapter focuses on Grad-CAM, Grad-CAM++, and NBDT. Before we proceed with the implementation, we take a deep dive into the following concepts.

Grad-CAM

Class activation maps (CAM) is a technique that extracts the heat maps that highlight the spatial information affecting the results. The CAM architecture is shown in Figures 10-1a and 10-1b.

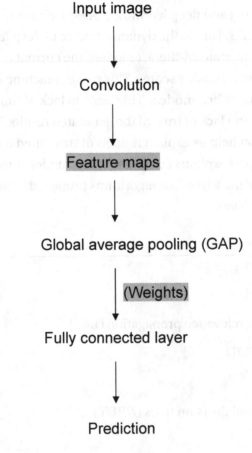

Figure 10-1a. *The CAM architecture*

The generated feature maps are passed through global average pooling to extract the weights. The weights are passed through a fully connected layer to output classification results.

The highlighted sections in the architecture, i.e. the feature maps and weights, are used to generate the heat maps for the predicted class.

$$\text{Weighted sum of feature maps} = \sum k \, (wk * Ak^{class})$$

Where k represents the feature maps from the last convolution layer.

Grad-CAM is similar to CAM until the feature maps generation step. After this step, any neural network can be added (such as VGG, ResNet, etc.), which can be differentiable to get back the gradients. Based on the prediction result, the gradients with respect to each feature map are calculated. Neuron importance (alpha values/ weights) are calculated for each feature map using "global average pooling" of gradients over the width and depth of feature map dimension (i,j). Grad-CAM architecture is shown in Figure 10-1b. The highlighted components are multiplied to generate heat maps.

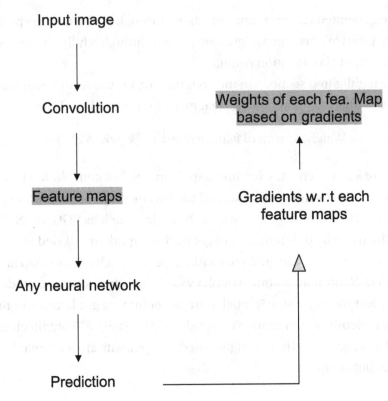

Figure 10-1b. *The CAM architecture*

Grad-CAM++

This is similar to the Grad-CAM algorithm but differs in the back-propagation step. In simple terms, Grad-CAM uses first order gradients during back-propagation. In Grad-CAM++, second order gradients are used, thus making the process more sophisticated.

In Grad-CAM, the feature maps with less spatial information are not given importance compared to the feature maps with high spatial information in the final heat maps. Images containing multiple objects or single objects can't be detected in heat maps, leading to low accuracy and explainability.

In Grad-CAM++, this issue is resolved by giving importance to all feature maps in the final heat maps. Figure 10-2 shows the overall architecture.

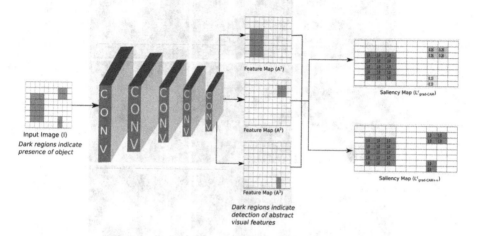

Figure 10-2. *The Grad-CAM++ architecture*

Figure 10-3 shows the difference in the outcomes from Grad-CAM and Grad-CAM++.

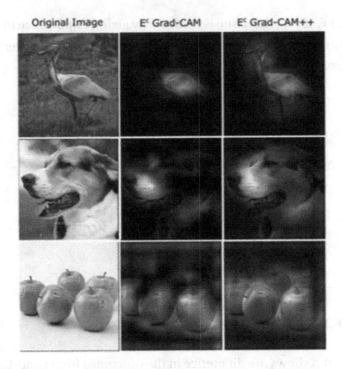

Figure 10-3. *Outcome from Grad-CAM and Grad-CAM++*

NBDT

This stands for neural based decision trees. Many algorithms are proposed for explainability of models, but the concept of results interpretability is missed.

- **Explainability:** Understanding the internal mechanics of a model, i.e., how it works internally.

- **Interpretability:** Understanding the cause and effect on the results, i.e., on what basis particular results are generated.

Decision trees are white-box models since they make it easy to understand how the nodes are split. This makes DT explainable.

It is also easy to know the predicted output with respect to changes in inputs. This makes DT interpretable.

The lagging point in DT is model accuracy when compared to deep learning models. NBDT has a built-in combination of decision trees (for explainability and interpretability) and neural networks (for accuracy). Figure 10-4 shows the NBDT flow.

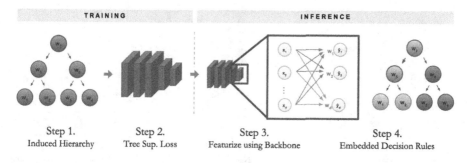

Figure 10-4. *NBDT Flow*

Step 1

This step trains a CNN model for image classification. Extract the weights for each class prediction (w1, w2,....), where w1 represents the hidden weights vector for predicting class 1.

The nearest vectors (also called leafs) are clustered to form the intermediate nodes. These intermediate nodes are clustered until the root node is reached. This hierarchy is called *induced hierarchy*. Following this, we converted NN to a DT.

The names for the intermediate nodes are extracted based on the WordNet module. (For example, dog and cat are leaf nodes. An intermediate node could be "animal," which is extracted from WordNet.)

Step 2

This step calculates the classification loss and fine tunes the model. Classification loss can be calculated using two modes.

- Hard mode

- Soft mode

The difference between these two modes is shown in Figure 10-5.

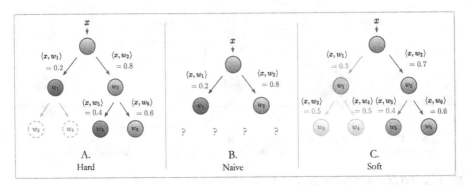

Figure 10-5. *Difference between hard mode and soft mode*

$$Loss\ (Total) = Loss\ (original) + Loss\ (hard\ or\ soft)$$

The hard or soft loss is added to the original loss (from CNN) to get the final loss.

Steps 3 and 4

Based on the final loss calculated, the model is fine-tuned and the decision tree (hierarchy) is updated.

Grad-CAM and Grad-CAM++ Implementation

First we discuss Grad-CAM and Grad-CAM++ implementation on a single image, then we cover NBDT implementation on a single image.

Grad-CAM and Grad-CAM++ Implementation on a Single Image

Step 1: Perform an image transformation on the input image (see Figure 10-6).

Figure 10-6. *Performing image transformation on the input image*

This includes:

- Resizing the image per the architecture in Step 2.

- Converting the resized image to a tensor for faster computation using PyTorch.

- Normalizing the image to make convergence faster
  during the training process. See Figures 10-7 and 10-8.

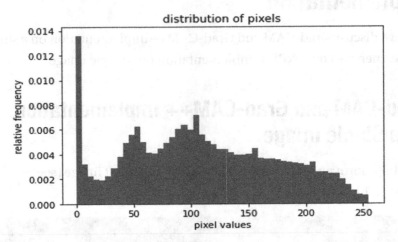

Figure 10-7. *Before transformation*

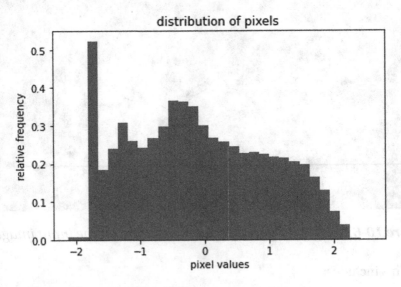

Figure 10-8. *After transformation*

```
#Transform input Image- Resize before passing to the model
resized_torch_img = transforms.Compose([transforms.Resize((224,
224)),transforms.ToTensor()])(pil_img).to(device)
```

```
#Image normalization
normalized_torch_img = transforms.Normalize([0.485, 0.456,
0.406], [0.229, 0.224, 0.225])(resized_torch_img)[None]
```

Step 2: Load the neural network architecture (with pretrained weights in this case). The following pretrained models are tested to compare their results:

- AlexNet

- VGG16

- ResNet101

- DenseNet161

- SqueezeNet

```
#Supported architectures in pytorch-gradcam library
model_alexnet = models.alexnet(pretrained=True)
model_vgg = models.vgg16(pretrained=True)
model_resnet = models.resnet101(pretrained=True)
model_densenet = models.densenet161(pretrained=True)
model_squeezenet = models.squeezenet1_1(pretrained=True)
```

Step 3: Select the feedback layer in the neural network to backpropagate the gradient. The preferred layer is selected for backpropagating the gradient.

```
#Storing the models as dictionary item with the respective
layers where gradients will be taken
loaded_configs = [
    dict(model_type='alexnet', arch=model_alexnet, layer_
    name='features_11'),
```

```
    dict(model_type='vgg', arch=model_vgg, layer_
    name='features_29'),
    dict(model_type='resnet', arch=model_resnet, layer_
    name='layer4'),
    dict(model_type='densenet', arch=model_densenet, layer_
    name='features_norm5'),
    dict(model_type='squeezenet', arch=model_squeezenet, layer_
    name='features_12_expand3x3_activation')]
```

Step 4: Load the Grad-CAM and Grad-CAM++ models.

Both models are loaded from the pytorch-gradcam library.

```
# Load the config to the "Grad CAM" and "Grad CAM ++"
# Only "Grad CAM" and "Grad CAM ++" available in this library
for model_config in loaded_configs:
    model_config['arch'].to(device).eval()

#Save "Grad CAM" and "Grad CAM ++" instances for all available
architectures(loaded_configs)
cams = [[cls.from_config(**model_config) for cls in (GradCAM,
GradCAMpp)] for model_config in loaded_configs]
```

Step 5: Pass the transformed input image and generate the heat map.

Pass the transformed image into both models and generate the results in the form of heat maps. These heat maps will highlight the key patches in the image.

```
#Load the normalized image to the "gradcam , gradcam ++"
function under each architecture to produce heatmaps and result
images = []
for gradcam, gradcam_pp in cams:
    mask, _ = gradcam(normalized_torch_img)
    heatmap, result = visualize_cam(mask, resized_torch_img)
```

```
mask_pp, _ = gradcam_pp(normalized_torch_img)
heatmap_pp, result_pp = visualize_cam(mask_pp, resized_
torch_img)

images.extend([resized_torch_img.cpu(),result,result_pp])
```

```
#Grid the original image, result from gradcam, result from
gradcam++
grid_image = make_grid(images, nrow=3)
```

Step 6: Combine the input image and heat map to visualize the important features selected for classification.

Convert the output tensor into the Python readable image to visualize the result. The following output is generated using DenseNet pretrained weights.

Input image>> Output from Grad-CAM >> Output from Grad-CAM++

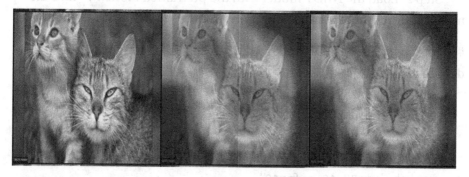

Figure 10-9. Output

NBDT Implementation on a Single Image

Step 1: Perform image transformation on the input image, similar to Grad-CAM.

```
#Function to Load image and perform image transformation
(resize,center crop,converting to tensor,normalization)
def load_image():
    assert len(sys.argv) > 1
    im = load_image_from_path("image_path")
    transform = transforms.Compose([
      transforms.Resize(32),
      transforms.CenterCrop(32),
      transforms.ToTensor(),
      transforms.Normalize((0.4914, 0.4822, 0.4465), (0.2023,
      0.1994, 0.2010)),
    ])
    x = transform(im)[None]
    return x
```

Step 2: Load the NBDT model with the pretrained model. To find the model hierarchy, the wordnet library is used.

```
#Function to load NBDT model with pretrained weights
def load_model():
    model = wrn28_10_cifar10()
    model = HardNBDT(
      pretrained=True,
      dataset='CIFAR10',
      arch='wrn28_10_cifar10',
      model=model)
    return model
```

Step 3: Predict the output from the HardNBDT model. Convert the predicted result and hierarchy into known classes.

```
#Function to output the classification result and hierarchy
def hierarchy_output(outputs, decisions):
    _, predicted = outputs.max(1)
    predicted_class = DATASET_TO_CLASSES['CIFAR10']
    [predicted[0]]
    print('Predicted Class:', predicted_class,
     '\n\nHierarchy:',
     ', '.join(['\n{} ({:.2f}%)'.format(info['name'],
    info['prob'] * 100)
        for info in decisions[0]][1:]))
```

Step 4: Output the predicted class and hierarchy from the decision tree.

```
def main():
    model = load_model()
    x = load_image()
    outputs, decisions = model.forward_with_decisions(x)
    hierarchy_output(outputs, decisions)

if __name__ == '__main__':
    main()
```

Here is the result:

```
Predicted Class: horse

Hierarchy:
animal (99.52%),
ungulate (98.52%),
horse (99.71%)
```

Summary

Explainability plays a major role going forward, because everyone wants to understand what is going on behind the scenes. Business leaders will have a hard time believing the AI models. Without being able to explain the results, all AI solutions are incomplete, and it's no different for computer vision.

Keeping that in mind, we went through the various libraries for explainability in this chapter. We learned about the concepts of CAM, Grad-CAM, and Grad-CAM++. Along with that, we implemented explainability using pretrained models and predictions. This has been a simple introduction to explainability; there are plenty of things to learn and implement.

Index

A

AlexNet, 335
  architecture, 24
  definition, 23
  inception architecture,
    30–33
  ResNet, 27–29
  VGG, 25–27
Anomaly detection, 7, 227,
    229–231, 259
argmax function, 60
Artificial intelligence, 1, 2
Autoencoder
  build network, 256
  dataset object, 255
  error metric score, 258
  output, 259
  reconstruction loss, 258
  training network,
    255, 257

B

Batch normalization,
    26, 31, 33, 34, 44, 72, 73,
    277, 278
Boosted Cascade, 86–90

C

Class activation maps (CAM), 326
Classification, 3, 4, 19, 36, 43–47, 49
Computer Vision (CV)
  applications, 3, 4, 6, 8
  channels, 9, 10
  deep learning model, 33
  definition, 2, 3
  models, 30, 37, 41, 43, 44, 323, 325
Convolutional neural network
    (CNN), 178
  AlexNet, 23
  feature map/receptive
    fields, 21–23
  feature maps, 14
  kernels, 12, 13
  machine learning or deep
    learning, 12
  padding, 14
  pooling, 17–20
  receptive field, 15, 17

D

Darknet framework, 123, 127
Data augmentation, 35, 37, 44,
    47, 71, 72

© Akshay Kulkarni, Adarsha Shivananda, and Nitin Ranjan Sharma 2022
A. Kulkarni et al., *Computer Vision Projects with PyTorch*,
https://doi.org/10.1007/978-1-4842-8273-1

W, X

Y, Z

Printed in the United States
by Baker & Taylor Publisher Services